Advanced Maths Essentials
Core 4 for AQA

1	Algebra and functions	1
	1.1 Simplifying rational expressions	1
	1.2 Algebraic division	3
	1.3 Partial fractions	7
2	Coordinate geometry in the (x, y) plane	13
	2.1 Parametric equations	13
3	Sequences and series	18
	3.1 Binomial series	18
4	Trigonometry	25
	4.1 Addition formulae	25
	4.2 Expressions for $a \cos \theta + b \sin \theta$	26
	4.3 Double angle formulae	31
5	Exponentials and logarithms	39
	5.1 Exponential growth and decay	39
6	Differentiation and integration	45
	6.1 Forming differential equations	45
	6.2 Solving differential equations	46
	6.3 Differentiation of implicit and parametric functions	50
	6.4 Integration using partial fractions	55
7	Vectors	63
	7.1 Vector geometry	63
	7.2 Position vectors and distance	69
	7.3 Vector equations of lines	71
	7.4 Scalar product	75
	Practice exam paper	84
	Answers	86

Welcome to Advanced Maths Essentials: Core 4 for AQA. This book will help you to improve your examination performance by focusing on the essential skills you will need in your AQA Core 4 examination. It has been divided by chapter into the main topics that need to be studied. Each chapter has then been divided by sub-headings, and the description below each sub-heading gives the AQA specification for that aspect of the topic.

The book contains scores of worked examples, each with clearly set-out steps to help solve the problem. You can then apply the steps to solve the Skills Check questions in the book and past exam questions at the end of each chapter. If you feel you need extra practice on any topic, you can try the Skills Check Extra exercises on the accompanying CD-ROM. At the back of this book there is a sample exam-style paper to help you test yourself before the big day.

Some of the questions in the book have a symbol next to them. These questions have a PowerPoint® solution (on the CD-ROM) that guides you through suggested steps in solving the problem and setting out your answer clearly.

Using the CD-ROM

To use the accompanying CD-ROM simply put the disc in your CD-ROM drive, and the menu should appear automatically. If it doesn't automatically run on your PC:

1. Select the My Computer icon on your desktop.
2. Select the CD-ROM drive icon.
3. Select Open.
4. Select core4_for _aqa.exe.

If you don't have PowerPoint® on your computer you can download PowerPoint 2003 Viewer®. This will allow you to view and print the presentations. Download the viewer from http://www.microsoft.com

Pearson Education Limited
Edinburgh Gate
Harlow
Essex
CM20 2JE
England
www.longman.co.uk

© Pearson Education Limited 2007

The rights of Janet Crawshaw and Kathryn Scott to be identified as the authors of this work have been asserted by them in accordance with the Copyright, Designs and Patents Act, 1988.

All rights reserved. No part of this publication may be reproduced, stored in a retrieval system, or transmitted in any form or by any means, electronic, mechanic, photocopying, recording, or otherwise without either the prior written permission of the Publishers or a licence permitting restricted copying in the United Kingdom issued by the Copyright Licensing Agency Ltd, 90 Tottenham Court Road, London W1P 9HE.

First published 2007
Second impression 2011
978-0-582-83686-0

Design by Ken Vail Graphic Design

Cover design by Raven Design

Typeset by Tech-Set, Gateshead

Printed in Malaysia (CTP-VP)

The publisher wishes to draw attention to the Single-User Licence Agreement at the back of the book.
Please read this agreement carefully before installing and using the CD-ROM.

AQA(AEB)/AQA examination materials are reproduced by permission of the Assessment and Qualifications Alliance. All such questions have a reference in the margin. AQA can accept no responsibility whatsoever for accuracy of any solutions or answers to these questions.

Every effort has been made to ensure that the structure and level of sample question papers matches the current specification requirements and that solutions are accurate. However, the publisher can accept no responsibility whatsoever for accuracy of any solutions or answers to these questions. Any such solutions or answers may not necessarily constitute all possible solutions.

1 Algebra and functions

1.1 Simplifying rational expressions

Rational functions. Simplification of rational expressions including factorising and cancelling.

To **simplify** an algebraic fraction:

- fully factorise the numerator and the denominator
- cancel common factors.

Example 1.1 Simplify $\dfrac{x^2 - 2x}{3x^2 - 7x + 2}$.

Step 1: Factorise both the numerator and the denominator fully.

Step 2: Cancel common factors.

$$\frac{x^2 - 2x}{3x^2 - 7x + 2} = \frac{x(x - 2)}{(3x - 1)(x - 2)}$$

$$= \frac{x(x - 2)^1}{(3x - 1)(x - 2)^1}$$

$$= \frac{x}{3x - 1}$$

Tip: **Do not** attempt any cancelling until the expressions are in factorised form.

Note: This *cannot* be simplified further.

To **multiply** algebraic fractions:

- fully factorise all numerators and denominators
- cancel common factors
- multiply the numerators and multiply the denominators.

To **divide** algebraic fractions:

- change $\div$ to $\times$ and invert the fraction immediately after the $\div$ sign
- follow the steps for multiplication.

Example 1.2 Express $\dfrac{4y^2 - 9}{2y^2 + 13y + 15} \div \dfrac{2y^2 - 3y}{y^2}$ as a single fraction in its simplest form.

Step 1: Change $\div$ to $\times$ and invert the second fraction.

Step 2: Factorise the numerators and denominators fully.

Step 3: Cancel common factors.

Step 4: Multiply the numerators and multiply the denominators.

$$\frac{4y^2 - 9}{2y^2 + 13y + 15} \div \frac{2y^2 - 3y}{y^2} = \frac{4y^2 - 9}{2y^2 + 13y + 15} \times \frac{y^2}{2y^2 - 3y}$$

$$= \frac{(2y + 3)(2y - 3)}{(2y + 3)(y + 5)} \times \frac{y^2}{y(2y - 3)}$$

$$= \frac{{}^1(2y + 3)(2y - 3)^1}{{}_1(2y + 3)(y + 5)} \times \frac{{}^1y \times y}{{}_1y(2y - 3)^1}$$

$$= \frac{y}{y + 5}$$

Recall: $a^2 - b^2 = (a - b)(a + b)$

Note: In practice, steps 2 and 3 would be done in one line.

To **add** or **subtract** algebraic fractions:

- factorise the denominators
- express each term as a fraction whose denominator is the lowest common multiple of all the denominators (lowest common denominator)
- add or subtract the fractions, writing a common denominator
- simplify the numerator, factorising if possible
- cancel common factors if possible.

1

Example 1.3 Express $\dfrac{5}{x-1} + \dfrac{3}{x+2}$ as a fraction in its simplest form.

Step 1: Write each term as a fraction with the lowest common denominator.

$$\dfrac{5}{x-1} + \dfrac{3}{x+2} = \dfrac{5(x+2)}{(x-1)(x+2)} + \dfrac{3(x-1)}{(x-1)(x+2)}$$

Step 2: Add or subtract the fractions.

$$= \dfrac{5(x+2) + 3(x-1)}{(x-1)(x+2)}$$

Step 3: Simplify the numerator.

$$= \dfrac{5x + 10 + 3x - 3}{(x-1)(x+2)}$$

$$= \dfrac{8x + 7}{(x-1)(x+2)}$$

Tip: The lowest common denominator is $(x-1)(x+2)$.

Tip: Do not expand the denominator; leave it in factorised form, as this is its simplest form.

Example 1.4
a Express $\dfrac{x}{x-4} - \dfrac{28}{x^2 - x - 12}$ as a fraction in its simplest form.

b Hence solve $\dfrac{x}{x-4} - \dfrac{28}{x^2 - x - 12} = 5$.

Step 1: Factorise the denominators.

a $\dfrac{x}{x-4} - \dfrac{28}{x^2 - x - 12} = \dfrac{x}{x-4} - \dfrac{28}{(x-4)(x+3)}$

Step 2: Write each term as a fraction with the lowest common denominator.

$$= \dfrac{x(x+3)}{(x-4)(x+3)} - \dfrac{28}{(x-4)(x+3)}$$

Step 3: Add or subtract the fractions.

$$= \dfrac{x(x+3) - 28}{(x-4)(x+3)}$$

Step 4: Simplify the numerator.

$$= \dfrac{x^2 + 3x - 28}{(x-4)(x+3)}$$

$$= \dfrac{(x-4)(x+7)}{(x-4)(x+3)}$$

Step 5: Cancel common factors, where possible.

$$= \dfrac{{}^1\cancel{(x-4)}(x+7)}{{}_1\cancel{(x-4)}(x+3)}$$

$$= \dfrac{x+7}{x+3}$$

Tip: Take care here. The lowest common denominator is $(x-4)(x+3)$, not $(x-4)^2(x+3)$.

Tip: Factorise the numerator if possible.

Step 1: Form an equation using the simplified form from a.

b $\dfrac{x}{x-4} - \dfrac{28}{x^2 - x - 12} = 5$

$\Rightarrow \dfrac{x+7}{x+3} = 5$

Step 2: Solve the equation.

$x + 7 = 5(x + 3)$

$x + 7 = 5x + 15$

$4x = -8$

$x = -2$

Tip: Eliminate the denominator by multiplying both sides of the equation by $(x+3)$.

1.2 Algebraic division

Algebraic division (using identities). The use of the Factor and Remainder Theorems for divisors of the form $(ax + b)$.

Algebraic division

Algebra fractions are either proper or improper.

An algebra fraction is **proper** if the degree of the denominator is greater than the degree of the numerator, for example $\dfrac{2x + 3}{x^2 + 7x + 12}$.

An algebra fraction is **improper** if the degree of the denominator is less than or equal to the degree of the numerator, for example

$$\dfrac{4x + 11}{x + 2} \text{ or } \dfrac{3x^2 - 4}{x + 2}.$$

You could be asked to divide a polynomial expression by a linear expression. You could do this using algebraic long division or by using identities.

Note: The degree of a polynomial is the highest power of x, for example $2x^3 + 4x + 1$ has degree 3.

Note: You are not required to know how to do algebraic long division, but you can use it if you wish (C1 Section 1.12).

Example 1.5 Express $\dfrac{4x + 11}{x + 2}$ in the form $A + \dfrac{B}{x + 2}$, where A and B are constants to be found.

Step 1: Write each term with a common denominator.
$$\dfrac{4x + 11}{x + 2} \equiv A + \dfrac{B}{x + 2}$$

$$\dfrac{4x + 11}{x + 2} \equiv \dfrac{A(x + 2)}{x + 2} + \dfrac{B}{x + 2}$$

Step 2: Add the fractions on the right-hand side.
$$\dfrac{4x + 11}{x + 2} \equiv \dfrac{A(x + 2) + B}{x + 2}$$

Step 3: Equate the numerators.
$$4x + 11 \equiv A(x + 2) + B$$

Step 4: Equate coefficients to find the values of A and B.
Coefficient of x: $4 = A$
Constant term: $11 = 2A + B \Rightarrow B = 3$

Step 5: Write the expression in the required format.
Hence $\dfrac{4x + 11}{x + 2} \equiv 4 + \dfrac{3}{x + 2}$.

Note: Using long division:
$$\begin{array}{r} 4 \\ x+2\overline{)4x + 11} \\ \underline{4x + 8} \\ 3 \end{array}$$
So $\dfrac{4x + 11}{x + 2} = 4 + \dfrac{3}{x + 2}$

Note: You could substitute values of x into the identity. For example, letting $x = -2$ makes the coefficient of A zero.

Note: $\dfrac{3}{x + 2}$ is a proper fraction.

Example 1.6 Express $\dfrac{2x^3 + 5x^2 + 3x - 1}{x + 1}$ in the form $Ax^2 + Bx + \dfrac{C}{x+1}$, where A, B and C are constants to be found.

Note: The given expression is an improper algebra fraction.

Step 1: Write each term with a common denominator.

$$\dfrac{2x^3 + 5x^2 + 3x - 1}{x + 1} \equiv Ax^2 + Bx + \dfrac{C}{x+1}$$

$$\dfrac{2x^3 + 5x^2 + 3x - 1}{x + 1} \equiv \dfrac{Ax^2(x+1)}{x+1} + \dfrac{Bx(x+1)}{x+1} + \dfrac{C}{x+1}$$

Note: Using long division:

$$\begin{array}{r} 2x^2 + 3x \\ x+1\overline{\smash{)}2x^3 + 5x^2 + 3x - 1} \\ \underline{2x^3 + 2x^2} \\ 3x^2 + 3x \\ \underline{3x^2 + 3x} \\ -1 \end{array}$$

$$\dfrac{2x^3 + 5x^2 + 3x - 1}{x+1} \equiv 2x^2 + 3x - \dfrac{1}{x+1}$$

Step 2: Add the fractions on the right-hand side.

$$\dfrac{2x^3 + 5x^2 + 3x - 1}{x + 1} \equiv \dfrac{Ax^2(x+1) + Bx(x+1) + C}{x + 1}$$

Step 3: Equate the numerators.

$$2x^3 + 5x^2 + 3x - 1 \equiv Ax^2(x+1) + Bx(x+1) + C$$

i.e. $2x^3 + 5x^2 + 3x - 1 \equiv Ax^3 + Ax^2 + Bx^2 + Bx + C$

Step 4: Equate coefficients to find the values of A, B and C.

Equating the coefficients of x^3: $\quad 2 = A$
Equating the coefficients of x^2: $\quad 5 = A + B \Rightarrow B = 3$
Equating the constant terms: $\quad -1 = C$

Step 5: Write the expression in the required format.

Therefore $\dfrac{2x^3 + 5x^2 + 3x - 1}{x + 1} \equiv 2x^2 + 3x + \dfrac{-1}{x+1}$

i.e. $\dfrac{2x^3 + 5x^2 + 3x - 1}{x + 1} \equiv 2x^2 + 3x - \dfrac{1}{x+1}$

Note: It is more usual to write $+\dfrac{-1}{x+1}$ in the form shown.

Example 1.7 Express $\dfrac{3x^2 - 10x - 19}{(x+1)(x-5)}$ in the form $A + \dfrac{Bx + C}{(x+1)(x-5)}$, where A, B and C are constants to be found.

Note: This is an improper algebra fraction.

Step 1: Write each term with a common denominator.

$$\dfrac{3x^2 - 10x - 19}{(x+1)(x-5)} \equiv A + \dfrac{Bx + C}{(x+1)(x-5)}$$

$$\dfrac{3x^2 - 10x - 19}{(x+1)(x-5)} \equiv \dfrac{A(x+1)(x-5)}{(x+1)(x-5)} + \dfrac{Bx + C}{(x+1)(x-5)}$$

Tip: The lowest common denominator is $(x+1)(x-5)$.

Step 2: Add the fractions on the right-hand side.

$$\dfrac{3x^2 - 10x - 19}{(x+1)(x-5)} \equiv \dfrac{A(x+1)(x-5) + Bx + C}{(x+1)(x-5)}$$

Step 3: Equate the numerators.

$3x^2 - 10x - 19 \equiv A(x+1)(x-5) + Bx + C$

i.e. $3x^2 - 10x - 19 \equiv A(x^2 - 4x - 5) + Bx + C$

Step 4: Equate coefficients to find the values of A, B and C.

Equating the coefficients of x^2: $\quad 3 = A$
Equating the coefficients of x: $\quad -10 = -4A + B \Rightarrow B = 2$
Equating the constant terms: $\quad -19 = -5A + C \Rightarrow C = -4$

Step 5: Write the expression in the required format.

Therefore $\dfrac{3x^2 - 10x - 19}{(x+1)(x-5)} \equiv 3 + \dfrac{2x - 4}{(x+1)(x-5)}$.

Note: $\dfrac{2x-4}{(x+1)(x-5)}$ can be expressed in partial fractions (Section 1.3).

Remainder Theorem and Factor Theorem

In *Core 1* you used the Remainder Theorem and the Factor Theorem for divisors of the form $(x + a)$. In *Core 4* this is extended to include divisors of the form $(ax + b)$.

Recall: Remainder and Factor Theorems (C1 Sections 1.10 and 1.11).

Remainder Theorem:

The remainder when $f(x)$ is divided by $(ax + b)$ is given by $f\left(-\dfrac{b}{a}\right)$.

Tip: Solve $ax + b = 0$ to find the value of x to substitute into $f(x)$.

Factor Theorem:

$f\left(-\dfrac{b}{a}\right) = 0 \Leftrightarrow ax + b$ is a factor of $f(x)$.

Note: When $ax + b$ is a factor of $f(x)$, the remainder is zero.

Example 1.8 It is given that $f(x) = 2x^2 + x - 6$. Find the remainder when $f(x)$ is divided by $2x - 5$.

Step 1: Substitute an appropriate value for x.

Now $2x - 5 = 0$ when $x = \tfrac{5}{2}$.
Substituting into $f(x)$ gives
$f(\tfrac{5}{2}) = 2(\tfrac{5}{2})^2 + \tfrac{5}{2} - 6 = 9$

Tip: Solving $2x + 5 = 0$ gives the value of x to substitute.

Step 2: Apply the Remainder Theorem.

Since $f(\tfrac{5}{2}) = 9$, the remainder is 9.

Example 1.9 It is given that $P(x) = 6x^3 + 13x^2 - 10x - 24$.

  a Show that $(2x + 3)$ is a factor of $P(x)$.

  b Write $P(x)$ in the form $(2x + 3)Q(x)$, where $Q(x)$ is a quadratic factor to be found.

  c Hence express $P(x)$ as the product of three linear factors and solve the equation $P(x) = 0$.

Step 1: Substitute an appropriate value for x.

a $P(x) = 6x^3 + 13x^2 - 10x - 24$
Now $2x + 3 = 0$ when $x = -\tfrac{3}{2}$.
Substituting into $P(x)$:
$P(-\tfrac{3}{2}) = 6(-\tfrac{3}{2})^3 + 13(-\tfrac{3}{2})^2 - 10(-\tfrac{3}{2}) - 24 = 0$

Step 2: Apply the Factor Theorem.

By the Factor Theorem,
since $P(-\tfrac{3}{2}) = 0$, $(2x + 3)$ is a factor of $P(x)$.

Step 1: Express $Q(x)$ in general form.

b $P(x) = (2x + 3)Q(x)$
$= (2x + 3)(ax^2 + bx + c)$

Step 2: Equate coefficients to find the values of a, b and c.

$6x^3 + 13x^2 - 10x - 24 \equiv (2x + 3)(ax^2 + bx + c)$
i.e. $6x^3 + 13x^2 - 10x - 24 \equiv 2ax^3 + 2bx^2 + 2cx + 3ax^2 + 3bx + 3c$

Equating coefficients:
  Coefficient of x^3: $\quad 6 = 2a \quad \Rightarrow a = 3$
  Coefficient of x^2: $\quad 13 = 2b + 3a \Rightarrow b = 2$
  Constant term: $\quad -24 = 3c \quad \Rightarrow c = -8$
So $Q(x) = 3x^2 + 2x - 8$

Step 1: Factorise the quadratic factor.

c $3x^2 + 2x - 8 = (3x - 4)(x + 2)$
So $P(x) = (2x + 3)(3x - 4)(x + 2)$

Step 2: Solve $P(x) = 0$.

$P(x) = 0 \Rightarrow (2x + 3)(3x - 4)(x + 2) = 0$
$\Rightarrow x = -\tfrac{3}{2}$ or $x = \tfrac{4}{3}$ or $x = -2$

SKILLS CHECK 1A: Simplifying rational expressions

1 Simplify fully:

a $\dfrac{x^2}{3x^2 - 2x}$

b $\dfrac{x^2 - 4}{x^2 + 3x - 10}$

c $\dfrac{4x + 12}{2x^2 - 18}$

d $\dfrac{3x^2 + 13x + 4}{2x^2 + 5x - 12}$

2 Express as a single fraction in its simplest form:

a $\dfrac{x^2 - 49}{x} \div \dfrac{x + 7}{x^2}$

b $\dfrac{10x + 15}{5x^2} \times \dfrac{x^2 + 4x}{2x + 3}$

c $\dfrac{x^3 - x}{4x^2} \times \dfrac{2x}{x + 1}$

d $\dfrac{4t^2 - 1}{2t^2 + 11t + 5} \div \dfrac{6t - 3}{t^2 + 5t}$

3 Express as a fraction in its simplest form:

a $\dfrac{5}{x - 1} + \dfrac{3}{x + 2}$

b $3 + \dfrac{2}{4 - y}$

c $\dfrac{1}{x - 2} - \dfrac{3}{x^2 - 7x + 10}$

d $\dfrac{2x}{x^2 - 2x - 3} + \dfrac{1}{x^2 - 1}$

4 a Simplify $\dfrac{x^3 + x^2 + x}{3x^2 - 2x} \times \dfrac{3x^2 + x - 2}{1 + x}$.

b Hence solve $\dfrac{x^3 + x^2 + x}{3x^2 - 2x} \times \dfrac{3x^2 + x - 2}{1 + x} = 3$.

5 Express $\dfrac{3x + 4}{x - 1}$ in the form $A + \dfrac{B}{x - 1}$, where A and B are integers to be found.

6 Express $\dfrac{3x^2 - 7x + 2}{3x - 4}$ in the form $Ax + B + \dfrac{C}{3x - 4}$, where A, B and C are integers to be found.

7 Given that $\dfrac{2x^3 - 3x^2 - 2x + 2}{x - 2} \equiv Ax^2 + Bx + \dfrac{C}{x - 2}$, find the values of A, B and C.

8 For the function $f(x) = 3x^3 + ax - 1$, the remainder when $f(x)$ is divided by $(2x - 1)$ is $\tfrac{3}{8}$.

a Find the value of a.

b Find the exact value of the remainder when $f(x)$ is divided by $(3x + 2)$.

9 It is given that $P(x) = 6x^3 - 5x^2 - 29x + 10$.

a Find the remainder when $P(x)$ is divided by $(3x - 1)$.

b Express $P(x)$ as the product of three linear factors.

c Hence solve the equation $P(x) = 0$.

SKILLS CHECK **1A EXTRA** is on the CD

1.3 Partial fractions

Partial fractions (denominators not more complicated than repeated linear denominators).

In Example 1.3 it was shown that

$$\frac{5}{x-1} + \frac{3}{x+2} = \frac{8x+7}{(x-1)(x+2)}$$

In the reverse of this process, a rational function is expressed as a sum of two or more simpler fractions called **partial fractions**.

You will need to be able to split the following proper and improper fractions into partial fractions:

- fractions with distinct linear factors in the denominator
- fractions with a repeated linear factor in the denominator.

Recall:
Algebraic fractions (Section 1.1).

Note:
The method of splitting a single fraction into partial fractions is also needed for work on the binomial series (Section 3.1) and integration (Section 6.4).

Fractions with distinct linear factors in the denominator

Notice that when $\frac{8x+7}{(x-1)(x+2)}$ is written in partial fractions as $\frac{5}{x-1} + \frac{3}{x+2}$, the denominators of the partial fractions are the factors of the original denominator. When splitting fractions of this type into partial fractions use this fact as a starting point.

Example 1.10 Express $\frac{2x-4}{x^2-2x-3}$ in partial fractions.

Step 1: Factorise the denominator.

$$\frac{2x-4}{x^2-2x-3} = \frac{2x-4}{(x-3)(x+1)}$$

Step 2: Set out the partial fractions using the factors of the denominator to make new denominators.
Step 3: Add the fractions.

Let $\frac{2x-4}{(x-3)(x+1)} \equiv \frac{A}{x-3} + \frac{B}{x+1}$

$$\frac{2x-4}{(x-3)(x+1)} \equiv \frac{A(x+1) + B(x-3)}{(x-3)(x+1)}$$

Step 4: Equate the numerators.

So
$$2x - 4 \equiv A(x+1) + B(x-3)$$

Step 5: To find A, substitute a value of x that will make the coefficient of B zero.

Substituting $x = 3$,
$$2 \times 3 - 4 = A(3+1) + B(3-3)$$
$$2 = 4A$$
$$A = \tfrac{1}{2}$$

Step 6: To find B, substitute a value of x that will make the coefficient of A zero.

Substituting $x = -1$,
$$2 \times -1 - 4 = A(-1+1) + B(-1-3)$$
$$-6 = -4B$$
$$B = \tfrac{3}{2}$$

Note:
Use the equivalence sign $\equiv$ because the identity is true for all values of x.
A and B are constants to be found.

Recall:
Adding fractions (Section 1.1).

Tip:
Substituting $x = 3$ makes the factor $(x-3)$ equal to zero.

Tip:
A common error is to state $A = 2$.

Tip:
Substituting $x = -1$ makes the factor $(x+1)$ equal to zero.

Step 7: Write out the partial fractions.

Therefore

$$\frac{2x-4}{x^2-2x-3} \equiv \frac{1}{2(x-3)} + \frac{3}{2(x+1)}$$

Note: $\frac{1}{2} \times \frac{1}{x-3} = \frac{1}{2(x-3)}$ and $\frac{3}{2} \times \frac{1}{x+1} = \frac{3}{2(x+1)}$.

In the above example the substitution method was used to find A and B. In the following example the method of equating coefficients is used. You will find it helpful to understand both methods when solving the more complicated examples later in the chapter.

Example 1.11 Express $\dfrac{5-6x}{(4x-3)(3x-2)}$ in partial fractions.

Step 1: Set out the partial fractions using the factors of the denominator to make new denominators.

Let $\dfrac{5-6x}{(4x-3)(3x-2)} \equiv \dfrac{A}{4x-3} + \dfrac{B}{3x-2}$

Step 2: Add the fractions.

$$\frac{5-6x}{(4x-3)(3x-2)} \equiv \frac{A(3x-2) + B(4x-3)}{(4x-3)(3x-2)}$$

Step 3: Equate the numerators.

So
$$5 - 6x \equiv A(3x-2) + B(4x-3)$$

Step 4: Expand the brackets.

$$5 - 6x \equiv 3Ax - 2A + 4Bx - 3B$$

Tip: Expanding the brackets will help you see the coefficients of the terms.

Step 5: Compare coefficients with the original numerator.

Equating coefficients of x,
$$-6 = 3A + 4B \qquad ①$$

Equating constant terms,
$$5 = -2A - 3B \qquad ②$$

Tip: Alternatively use the substitution method with $x = \frac{3}{4}$ and $x = \frac{2}{3}$. This will involve working with fractions accurately.

Step 6: Solve the simultaneous equations for A and B.

$① \times 2 \quad -12 = 6A + 8B \qquad ③$
$② \times 3 \quad 15 = -6A - 9B \qquad ④$
$③ + ④ \quad 3 = -B$
$\qquad\qquad B = -3$

Substituting back into ①,
$$-6 = 3A - 12$$
$$A = 2$$

Step 7: Write out the partial fractions.

Therefore

$$\frac{5-6x}{(4x-3)(3x-2)} \equiv \frac{2}{4x-3} - \frac{3}{3x-2}$$

Tip: $+\dfrac{-3}{3x-2}$ is the same as $-\dfrac{3}{3x-2}$.

Example 1.12
a Express $\dfrac{3x^2 + 7x + 8}{(x+1)(2x+1)(x-3)}$ in partial fractions.

b Hence differentiate $y = \dfrac{3x^2 + 7x + 8}{(x+1)(2x+1)(x-3)}$ with respect to x.

Step 1: Set out the partial fractions using the factors of the denominator to make new denominators.

a Let $\dfrac{3x^2 + 7x + 8}{(x+1)(2x+1)(x-3)} \equiv \dfrac{A}{x+1} + \dfrac{B}{2x+1} + \dfrac{C}{x-3}$

Step 2: Add the fractions.

$$\frac{3x^2 + 7x + 8}{(x+1)(2x+1)(x-3)} \equiv \frac{A(2x+1)(x-3) + B(x+1)(x-3) + C(x+1)(2x+1)}{(x+1)(2x+1)(x-3)}$$

Tip: There are three factors in the original denominator, so there will be three partial fractions.

So

Step 3: Equate the numerators.

$$3x^2 + 7x + 8 \equiv A(2x+1)(x-3) + B(x+1)(x-3) + C(x+1)(2x+1)$$

Step 4: To find A, substitute a value of x that will make the coefficients of B and C zero.

Substituting $x = -1$,
$$3(-1)^2 + 7(-1) + 8 = A(2 \times -1 + 1)(-1 - 3) + B \times 0 + C \times 0$$
$$4 = 4A$$
$$A = 1$$

Tip: Substitute $x = -1$ because then the factor $(x + 1)$ will equal zero.

Step 5: To find C, substitute a value of x that will make the coefficients of A and B zero.

Substituting $x = 3$,
$$3(3)^2 + 7 \times 3 + 8 = A \times 0 + B \times 0 + C(3 + 1)(2 \times 3 + 1)$$
$$56 = 28C$$
$$C = 2$$

Tip: Substitute $x = 3$ because then the factor $(x - 3)$ will equal zero.

Step 6: To find B, compare coefficients with the original numerator.

Equating coefficients of x^2,
$$3x^2 + \ldots \equiv A(2x^2 + \ldots) + B(x^2 + \ldots) + C(2x^2 + \ldots)$$
so $$3 = 2A + B + 2C$$
Substituting $A = 1$ and $C = 2$ gives
$$3 = 2 + B + 4$$
$$B = -3$$

Tip: You do not need to expand all the brackets, just the terms in x^2.

Tip: Alternatively you could substitute $x = -\frac{1}{2}$.

Step 7: Write out the partial fractions.

Therefore
$$\frac{3x^2 + 7x + 8}{(x + 1)(2x + 1)(x - 3)} = \frac{1}{x + 1} - \frac{3}{2x + 1} + \frac{2}{x - 3}$$

Tip: Check your answer by adding the partial fractions.

Step 1: Write your solution from **a** into a suitable format for differentiating.

b $y = \dfrac{3x^2 + 7x + 8}{(x + 1)(2x + 1)(x - 3)} = \dfrac{1}{x + 1} - \dfrac{3}{2x + 1} + \dfrac{2}{x - 3}$

$= (x + 1)^{-1} - 3(2x + 1)^{-1} + 2(x - 3)^{-1}$

Step 2: Differentiate with respect to x.

$\dfrac{dy}{dx} = -(x + 1)^{-2} + 6(2x + 1)^{-2} - 2(x - 3)^{-2}$

$= \dfrac{6}{(2x + 1)^2} - \dfrac{1}{(x + 1)^2} - \dfrac{2}{(x - 3)^2}$

Recall: Use the chain rule to differentiate $-3(2x + 1)^{-1}$ (C3 Section 4.2).

Fractions with a repeated linear factor in the denominator

A fraction such as $\dfrac{4x + 1}{x(2x - 1)^2}$ has three factors in the denominator so it will split into three partial fractions. These will be of the form $\dfrac{A}{x} + \dfrac{B}{2x - 1} + \dfrac{C}{(2x - 1)^2}$.

Note: If you have been taught the 'cover up' method, it does not work with fractions in this format.

Example 1.13 Express $\dfrac{4x + 1}{x(2x - 1)^2}$ in partial fractions.

Step 1: Set out the partial fractions using the factors of the denominator to make new denominators.

Let $\dfrac{4x + 1}{x(2x - 1)^2} \equiv \dfrac{A}{x} + \dfrac{B}{2x - 1} + \dfrac{C}{(2x - 1)^2}$

$\dfrac{4x + 1}{x(2x - 1)^2} \equiv \dfrac{A(2x - 1)^2}{x(2x - 1)^2} + \dfrac{Bx(2x - 1)}{x(2x - 1)^2} + \dfrac{Cx}{x(2x - 1)^2}$

Tip: Don't expand the brackets here; it's unnecessary and will make the calculations harder when substituting.

Step 2: Add the fractions.

$\dfrac{4x + 1}{x(2x - 1)^2} \equiv \dfrac{A(2x - 1)^2 + Bx(2x - 1) + Cx}{x(2x - 1)^2}$

Step 3: Equate the numerators.

So $4x + 1 \equiv A(2x - 1)^2 + Bx(2x - 1) + Cx$

Step 4: To find A, substitute a value of x that will make the coefficients of B and C zero.

Substituting $x = 0$,
$$4 \times 0 + 1 = A(2 \times 0 - 1)^2 + B \times 0 + C \times 0$$
$$A = 1$$

Tip: One of the factors is x so make this zero.

Step 5: To find C, substitute a value of x that will make the coefficients of A and B zero.

Substituting $x = \frac{1}{2}$,
$$4 \times \tfrac{1}{2} + 1 = A \times 0 + B \times 0 + C \times \tfrac{1}{2}$$
$$3 = \tfrac{1}{2}C$$
$$C = 6$$

Tip: Substitute $x = \frac{1}{2}$ because then the factor $(2x - 1)$ equals zero.

Step 6: To find B, compare coefficients with the original numerator.

Equating coefficients of x^2,
$$0 = 4A + 2B$$
$$0 = 4 + 2B$$
$$B = -2$$

Tip: Alternatively you could substitute a third value of x or equate coefficients of x or constant terms.

Step 7: Write out the partial fractions.

Therefore $\dfrac{4x + 1}{x(2x - 1)^2} = \dfrac{1}{x} - \dfrac{2}{2x - 1} + \dfrac{6}{(2x - 1)^2}$

When dealing with **improper fractions**, you will need to perform algebraic division. This is illustrated in the following example.

Example 1.14 $f(x) = \dfrac{2x^2 + 4x + 5}{(x - 2)(x + 1)}$.

Express $f(x)$ in the form $A + \dfrac{B}{x - 2} + \dfrac{C}{x + 1}$, where A, B and C are constants to be found.

Note: This is an improper fraction since the degree of both the numerator and denominator is 2.

Step 1: Write each term with a common denominator.

$$\dfrac{2x^2 + 4x + 5}{(x - 2)(x + 1)} \equiv A + \dfrac{B}{x - 2} + \dfrac{C}{x + 1}$$

$$\dfrac{2x^2 + 4x + 5}{(x - 2)(x + 1)} \equiv \dfrac{A(x - 2)(x + 1)}{(x - 2)(x + 1)} + \dfrac{B(x + 1)}{(x - 2)(x + 1)} + \dfrac{C(x - 2)}{(x - 2)(x + 1)}$$

Tip: The lowest common denominator is $(x - 2)(x + 1)$.

Step 2: Add the fractions on the right-hand side.

$$\dfrac{2x^2 + 4x + 5}{(x - 2)(x + 1)} \equiv \dfrac{A(x - 2)(x + 1) + B(x + 1) + C(x - 2)}{(x - 2)(x + 1)}$$

Step 3: Equate the numerators.
$$\Rightarrow 2x^2 + 4x + 5 \equiv A(x - 2)(x + 1) + B(x + 1) + C(x - 2)$$

Step 4: Equate coefficients to find A.

Equating the coefficients of x^2: $2 = A$

Step 5: To find B and C, substitute appropriate values of x.

Substituting $x = 2$,
$$2 \times 2^2 + 4 \times 2 + 5 = A \times 0 + B \times 3 + C \times 0$$
$$21 = 3B$$
$$B = 7$$

Substituting $x = -1$,
$$2 \times (-1)^2 + 4 \times (-1) + 5 = A \times 0 + B \times 0 + C \times (-3)$$
$$3 = -3C$$
$$C = -1$$

Note: You could perform algebraic long division to get
$$\dfrac{2x^2 + 4x + 5}{(x - 2)(x + 1)} \equiv 2 + \dfrac{6x + 9}{(x - 2)(x + 1)}$$
and then express $\dfrac{6x + 9}{(x - 2)(x + 1)}$ in partial fractions.

Step 6: Write out the answer in the required format.

Therefore $\dfrac{2x^2 + 4x + 5}{(x - 2)(x + 1)} \equiv 2 + \dfrac{7}{x - 2} - \dfrac{1}{x + 1}$

SKILLS CHECK 1B: Partial fractions

1 Express in partial fractions:

   a $\dfrac{4}{(x-3)(x+1)}$
   b $\dfrac{x-1}{(3x-5)(x-3)}$
   c $\dfrac{4x-13}{2x^2+x-6}$

2 The following fractions have repeated linear terms in their denominators. Express them in partial fractions.

   a $\dfrac{1}{x^2(x-1)}$
   b $\dfrac{2x^2+3}{(x+2)(x+1)^2}$
   c $\dfrac{4x^2+5x+9}{(2x-1)(x+2)^2}$

3 Express in partial fractions:

   a $\dfrac{x+27}{x^2-9}$
   b $\dfrac{3x}{(1-x)(1+x)^2}$

4 **a** Express $\dfrac{3}{(1-2x)(x+1)}$ in partial fractions.

   b Given the function $f(x) = \dfrac{3}{(1-2x)(x+1)}$, $x \neq -1$, $x \neq \tfrac{1}{2}$, find the coordinates of the minimum point of the curve $y = f(x)$.

5 **a** Express $\dfrac{5x-1}{1-x^2}$ in partial fractions.

   b Given that $f(x) = \dfrac{5x-1}{1-x^2}$, $x \neq \pm 1$, find

   i $f'(x)$
   ii $f''(0)$.

6 The function f is given by
$$f(x) = \dfrac{7-4x}{(2x+3)(x-5)}, \quad x \neq -\tfrac{3}{2}, x \neq 5$$

   a Express $f(x)$ in partial fractions.

   b Hence, or otherwise, prove that $f'(x) > 0$ for all values of x in the domain.

7 Express $\dfrac{11-5x^2}{(2+x)(1-x)}$ in the form $A + \dfrac{B}{2+x} + \dfrac{C}{1-x}$, where A, B and C are constants to be found.

8 Express $\dfrac{x^2-2x}{(x-4)(x-6)}$ in the form $A + \dfrac{B}{x-4} + \dfrac{C}{x-6}$, where A, B and C are constants to be found.

SKILLS CHECK 1B EXTRA is on the CD

Examination practice 1: Algebra and functions

There are further questions involving partial fractions in Examination practice 3 and Examination practice 6.

1. Express $\dfrac{3x^2 + 15x}{(2x + 1)^2} \div \dfrac{x^2 - 25}{2x^2 - 9x - 5}$ as a single fraction in its simplest form.

2.
   a. Express $\dfrac{13}{x^2 - 5x - 24} - \dfrac{1}{x + 3}$ as a fraction in its simplest form.

   b. Hence solve $\dfrac{13}{x^2 - 5x - 24} - \dfrac{1}{x + 3} = 1$.

3. a. Factorise fully the expression $x^3 - x$.

   b. Express $\dfrac{4x - 2}{x^3 - x}$ in partial fractions.

4. It is given that $y = \dfrac{7}{(2x - 1)(x + 3)}$.

   a. Express $\dfrac{7}{(2x - 1)(x + 3)}$ in partial fractions.

   b. Hence find i $\dfrac{dy}{dx}$ ii $\dfrac{d^2y}{dx^2}$.

5. Express $\dfrac{x(2x - 5)}{(x + 2)(x - 1)^2}$ in the form $\dfrac{A}{x + 2} + \dfrac{B}{x - 1} + \dfrac{C}{(x - 1)^2}$, where A, B and C are integers to be found.

6. Express $\dfrac{6(x^2 + x - 1)}{(x - 1)(2x + 1)}$ in the form $A + \dfrac{B}{x - 1} + \dfrac{C}{2x + 1}$, where A, B and C are integers to be found.

7. Show that $\dfrac{x^3 + 3x^2 + 10}{x^2 + 5x + 4}$ can be expressed in the form $A + Bx + \dfrac{C}{x + 1} + \dfrac{D}{x + 4}$, where A, B, C and D are constants to be found.

8. The polynomial f(x) is defined by
   $$f(x) = 2x^3 + x^2 - 13x + 6$$
   a. Find the remainder when f(x) is divided by
      i $2x + 1$ ii $2x - 1$

   b. Express f(x) as the product of three linear factors.

   c. Solve the equation f(x) = 0.

9. a. Use the Factor Theorem to show that $(3x - 1)$ is a factor of the polynomial $f(x) = 3x^3 + 5x^2 + 4x - 2$.

   b. Hence factorise $3x^3 + 5x^2 + 4x - 2$ as the product of a linear and a quadratic factor.

   c. Hence show that the equation f(x) = 0 has only one real solution and state the value of x for which f(x) = 0.

2 Coordinate geometry in the (x, y) plane

2.1 Parametric equations

Cartesian and parametric equations of curves and conversion between the two forms.

Up to now you have worked with Cartesian equations connecting two variables, usually x and y. Sometimes it is easier or more convenient to express x and y in terms of a third variable called a **parameter**.

For example,

$$x = t^2 + 1$$
$$y = t + 2$$

are the parametric equations of a curve.

Note: Here the parameter is t.

Example 2.1
a Draw the curve given by the parametric equations $x = t^2 + 1$, $y = t + 2$ for $-4 \leqslant t \leqslant 4$.
b Find a Cartesian equation of the curve.

Step 1: Draw up a table of values for x and y by substituting different values of t into x and y in the given interval.

a
t	-4	-3	-2	-1	0	1	2	3	4
$x = t^2 + 1$	17	10	5	2	1	2	5	10	17
$y = t + 2$	-2	-1	0	1	2	3	4	5	6

Tip: Use the patterns in the table to spot any errors.

Step 2: Plot the coordinates.

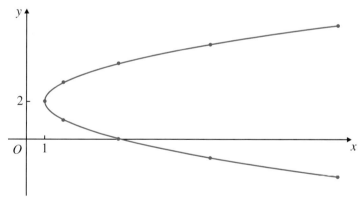

Tip: Each value of t gives a pair of coordinates, e.g. when $t = 2$, $x = 5$ and $y = 4$, so plot the point (5, 4).

Step 1: Rearrange one of the equations to find an expression for t.

b $x = t^2 + 1$, $y = t + 2$

$$y = t + 2$$

so

$$t = y - 2 \quad \text{①}$$

Since

$$x = t^2 + 1 \quad \text{②}$$

Step 2: Substitute the expression for t into the other parametric equation to eliminate t.

substituting ① into ② gives

$$x = (y - 2)^2 + 1$$

So a Cartesian equation of the curve is $x = (y - 2)^2 + 1$.

Tip: In this example it is easier to find an expression for t in terms of y than in terms of x.

Tip: Make sure your solution is in terms of x and y only.

Note: It is not always this straightforward to find the Cartesian form (see Examples 2.2 and 2.3).

Example 2.2 A curve has parametric equations

$$x = t + \frac{3}{t^2}, \quad y = t - \frac{3}{t^2}, \quad \text{where } t \neq 0$$

a Express $x + y$ and $x - y$ in terms of the parameter t.

b Hence verify that the Cartesian equation of the curve is

$$(x + y)^2(x - y) = 24$$

a $x = t + \dfrac{3}{t^2}$ ①

$y = t - \dfrac{3}{t^2}$ ②

Step 1: Add the equations. ① + ② $x + y = 2t$ ③

Step 2: Subtract the equations. ① − ② $x - y = \dfrac{6}{t^2}$ ④

Step 3: Eliminate t. **b** From ④, $t^2 = \dfrac{6}{x - y}$

From ③, $t = \tfrac{1}{2}(x + y) \Rightarrow t^2 = [\tfrac{1}{2}(x + y)]^2 = \tfrac{1}{4}(x + y)^2$

So $\tfrac{1}{4}(x + y)^2 = \dfrac{6}{x - y}$

$(x + y)^2(x - y) = 24$

Hence the Cartesian equation of the curve is $(x + y)^2(x - y) = 24$.

Tip: Find two expressions for t^2, then equate.

Curves with trigonometric parametric equations

To convert trigonometric parametric equations to Cartesian form, as before, you need to eliminate the parameter.

The identity $\cos^2 \theta + \sin^2 \theta \equiv 1$ is often useful here.

Recall: Trigonometric identities (C2 Section 3.6, C3 Section 2.3).

Example 2.3 Find the Cartesian equation of the curve given by the parametric equations

$$x = 2 \sin \theta - 3, \quad y = \cos \theta + 1$$

Step 1: Rearrange the equations making $\sin \theta$ and $\cos \theta$ the subject.

$x = 2 \sin \theta - 3$ $\qquad$ $y = \cos \theta + 1$

$2 \sin \theta = x + 3$ $\qquad$ $\cos \theta = y - 1$

$\sin \theta = \dfrac{x + 3}{2}$

Note: The parameter is θ.

Step 2: Eliminate θ by using an appropriate trig identity.

Since $\cos^2 \theta + \sin^2 \theta = 1$,

$$(y - 1)^2 + \frac{(x + 3)^2}{4} = 1$$

The Cartesian equation of the curve is $(y - 1)^2 + \dfrac{(x + 3)^2}{4} = 1$.

Tip:
$\cos \theta = y - 1$, so $\cos^2 \theta = (y - 1)^2$.
$\sin \theta = \dfrac{x + 3}{2}$, so $\sin^2 \theta = \left(\dfrac{x + 3}{2}\right)^2 = \dfrac{(x + 3)^2}{4}$.

Note: Try plotting this on a graphical calculator; the curve is an ellipse.

Example 2.4 A curve has parametric equations $x = 2t + 1$, $y = t^2 - 1$.

a Find the coordinates of the points of intersection of the curve with the axes.

b Find the points of intersection of the curve with the line $2y = x + 9$.

Step 1: Substitute $x = 0$ to find the value of t at the point where the curve cuts the y-axis.

a $x = 2t + 1$

When $x = 0$,
$$0 = 2t + 1$$
$$2t = -1$$
$$t = -\tfrac{1}{2}$$

Step 2: Substitute this value of t to find the y-coordinate.

When $t = -\tfrac{1}{2}$,
$$y = t^2 - 1 = (-\tfrac{1}{2})^2 - 1 = -\tfrac{3}{4}$$

So the curve cuts the y-axis at $(0, -\tfrac{3}{4})$.

$$y = t^2 - 1$$

Step 3: Substitute $y = 0$ to find the value of t at the point where the curve cuts the x-axis.

When $y = 0$,
$$0 = t^2 - 1$$
$$t^2 = 1$$
$$t = \pm 1$$

Tip: There will be two solutions when you take the square root.

Step 4: Substitute the values of t to find the x-coordinates.

When $t = 1$,
$$x = 2t + 1 = 2 \times 1 + 1 = 3$$
When $t = -1$,
$$x = 2t + 1 = 2 \times -1 + 1 = -1$$

So the curve cuts the x-axis at $(3, 0)$ and $(-1, 0)$.

Tip: Be careful to remember which coordinate you are looking for and substitute into the appropriate equation.

Step 1: Solve the equations of the line and the curve simultaneously by substituting both parametric equations into the equation of the line.

b $2y = x + 9$

At the point of intersection,
$$2(t^2 - 1) = (2t + 1) + 9$$
$$2t^2 - 2 = 2t + 10$$
$$2t^2 - 2t - 12 = 0$$
$$t^2 - t - 6 = 0$$
$$(t - 3)(t + 2) = 0$$
$$t = 3 \text{ or } -2$$

Tip: You are forming a quadratic equation in t. As usual you will need to collect terms and factorise to solve it.

Step 2: Substitute the values of t into the parametric equations to find the coordinates.

When $t = 3$,
$$x = 2t + 1 = 2 \times 3 + 1 = 7$$
$$y = t^2 - 1 = 3^2 - 1 = 8$$

When $t = -2$,
$$x = 2t + 1 = 2 \times -2 + 1 = -3$$
$$y = t^2 - 1 = (-2)^2 - 1 = 3$$

The points of intersection are $(7, 8)$ and $(-3, 3)$.

For examples involving tangents, normals and stationary points, see Section 6.3.

SKILLS CHECK 2A: Coordinate geometry in the (x, y) plane

1. For each of the following curves, given in parametric form, find the coordinates of any points where the curve cuts the y-axis:

   a $x = t - 1, y = t^2 + 1$

   b $x = 3\cos\theta, y = 4\sin\theta, 0 \leqslant \theta < 2\pi$

2. For each of the following curves, given in parametric form, find the coordinates of any points where the curve cuts the x-axis:

   a $x = 2t, y = 3 - t$

   b $x = 3t + 2, y = t^2 - 1$

3. A curve has parametric equations $x = \dfrac{at^2}{2}, y = a(t - 1)$, where a is a constant. Given that the curve passes through the point (2, 0), find the value of a.

4. Find a Cartesian equation for each of the following curves defined by the given parametric equations:

   a $x = t^2, y = 2t$

   b $x = a\cos\theta, y = b\sin\theta$

   c $x = t + 3, y = \dfrac{1}{t}, t \neq 0$

   d $x = \cos\theta, y = \tan\theta$

5. A curve is defined by the parametric equations

   $$x = \frac{3}{1 + t^2} \quad y = \frac{t}{1 + t^2}$$

   Find the coordinates of the point of intersection of the curve with the line $x + y = 1$.

6. Find the Cartesian equation of the curve defined by $x = \sec\theta, y = 3\tan\theta$.

7. **a** Find the Cartesian equation of the circle with parametric equations

   $x = 3\cos\theta + 2, \quad y = 3\sin\theta - 4$

   b Find the radius and the coordinates of the centre of the circle.

8. A curve is defined by the parametric equations

   $$x = t + \frac{1}{t}, \quad y = t - \frac{1}{t}, \quad t \neq 0$$

   a Express $(x + y)$ and $(x - y)$ in terms of t.

   b Hence find the Cartesian equation of the curve.

SKILLS CHECK 2A EXTRA is on the CD

Examination practice 2: Coordinate geometry in the (x, y) plane

There are further questions on curves defined by parametric equations in Examination practice 6.

1. A curve has parametric equations

   $$x = ct, \quad y = \frac{c}{t}, \quad t > 0, \quad \text{where } c \text{ is a positive constant}$$

   Given that the curve passes through the point (9, 1), find the value of c.

2 a Find a Cartesian equation for the curve defined by the parametric equations
$$x = t - 1, \quad y = t^3 + t^2$$
b Find the coordinates of the point at which the curve crosses the x-axis.

3 A curve is defined by the parametric equations
$$x = 2kt^2, \quad y = k(8t - t^4), \quad \text{where } k \text{ is a positive constant}$$
Given that the curve passes through the point $(24, 0)$, find the value of k.

4 A curve is defined by the parametric equations
$$x = 4\cos t - 1, \quad y = 4\sin t + 5$$
a By expressing the Cartesian equation of the curve in the form
$$(x - a)^2 + (y - b)^2 = r^2$$
show that the curve is a circle.

b Find the radius and the coordinates of the centre of the circle.

 5 A curve is defined by the parametric equations
$$x = 2t^2, \quad y = 4t - 1$$
Find the equation of the line joining the points on the curve where $t = -2$ and $t = 1$.

3 Sequences and series

3.1 Binomial series

Binomial series for any rational n.

The binomial expansion of $(1 + x)^n$, where n is a rational number, is given by

$$(1 + x)^n = 1 + \binom{n}{1}x + \binom{n}{2}x^2 + \binom{n}{3}x^3 + \cdots + \binom{n}{r}x^r + \cdots$$

$$= 1 + nx + \frac{n(n-1)}{2!}x^2 + \frac{n(n-1)(n-2)}{3!}x^3 + \cdots$$

$$+ \frac{n(n-1)\ldots(n-r+1)}{r!}x^r + \cdots$$

When n is a positive integer, the series is finite, with last term x^n. The series gives $(1 + x)^n$ exactly.

When n is not a positive integer, the series is infinite, since none of the factors of the coefficients will ever be zero.
Provided $-1 < x < 1$, the series is convergent and is a valid approximation to $(1 + x)^n$.

Tip: $\binom{n}{2}$ means $\frac{n(n-1)}{2!}$. Don't confuse it with $\frac{n}{2}$.

Recall: $r! = r(r-1)(r-2) \ldots \times 3 \times 2 \times 1$

Recall: Binomial expansion (C2 Section 2.5).

Note: $-1 < x < 1$ can be written $|x| < 1$ (C3 Section 1.4).

Example 3.1 Find the binomial expansion of the following in ascending powers of x up to and including the term in x^3. State the range of values of x for which each expansion is valid.

a $(1 + x)^{-3}$ **b** $\sqrt{1 - 2x}$ **c** $\dfrac{1}{(1 + 3x)^{\frac{2}{3}}}$

Step 1: Substitute into the binomial expansion.
Step 2: Simplify the terms.
Step 3: State when the expansion is valid.

a $(1 + x)^{-3} = 1 + (-3)x + \dfrac{(-3)(-4)}{2!}x^2 + \dfrac{(-3)(-4)(-5)}{3!}x^3 + \cdots$

$= 1 - 3x + 6x^2 - 10x^3 + \cdots$

The expansion is valid when $|x| < 1$.

Tip: Replace n by -3 in the expansion.

Note: The coefficients are $1, -3, 6, -10, \ldots$

Step 1: Write in index form.

b $\sqrt{1 - 2x} = (1 - 2x)^{\frac{1}{2}}$

Step 2: Substitute into the binomial expansion.

$= 1 + (\tfrac{1}{2})(-2x) + \dfrac{(\tfrac{1}{2})(-\tfrac{1}{2})}{2!}(-2x)^2 + \dfrac{(\tfrac{1}{2})(-\tfrac{1}{2})(-\tfrac{3}{2})}{3!}(-2x)^3 + \cdots$

Step 3: Simplify the terms.

$= 1 + (\tfrac{1}{2})(-2x) + \dfrac{(\tfrac{1}{2})(-\tfrac{1}{2})}{2!}(-2x)(-2x)$

$+ \dfrac{(\tfrac{1}{2})(-\tfrac{1}{2})(-\tfrac{3}{2})}{3!}(-2x)(-2x)(-2x) + \cdots$

$= 1 + (\tfrac{1}{2})(-2x) + \dfrac{(\tfrac{1}{2})(-\tfrac{1}{2})}{2!}(4x^2) + \dfrac{(\tfrac{1}{2})(-\tfrac{1}{2})(-\tfrac{3}{2})}{3!}(-8x^3) + \cdots$

$= 1 - x - \tfrac{1}{2}x^2 - \tfrac{1}{2}x^3 + \cdots$

Step 4: State when the expansion is valid.

The expansion is valid when $|-2x| < 1$, i.e. when $|x| < \tfrac{1}{2}$.

Tip: Replace n by $\tfrac{1}{2}$ and x by $-2x$ in the expansion.

Tip: Don't forget to square the -2 as well as the x: $(-2x)^2 = 4x^2$ not $-2x^2$. It can help to write this out in full.

Note: The coefficients are $1, -1, -\tfrac{1}{2}, -\tfrac{1}{2}, \ldots$

Tip: $-1 < -2x < 1$, so $\tfrac{1}{2} > x > -\tfrac{1}{2}$ i.e. $-\tfrac{1}{2} < x < \tfrac{1}{2}$ so $|x| < \tfrac{1}{2}$

Step 1: Write in index form.

c $\dfrac{1}{(1+3x)^{\frac{2}{3}}} = (1+3x)^{-\frac{2}{3}}$

Tip: Replace n by $-\frac{2}{3}$ and x by $3x$ in the expansion.

Step 2: Substitute into the binomial expansion.

$= 1 + (-\tfrac{2}{3})(3x) + \dfrac{(-\frac{2}{3})(-\frac{5}{3})}{2!}(3x)^2 + \dfrac{(-\frac{2}{3})(-\frac{5}{3})(-\frac{8}{3})}{3!}(3x)^3 + \cdots$

Step 3: Simplify the terms.

$= 1 + (-\tfrac{2}{3})(3x) + \dfrac{(-\frac{2}{3})(-\frac{5}{3})}{2!}(3x)(3x)$

$\quad + \dfrac{(-\frac{2}{3})(-\frac{5}{3})(-\frac{8}{3})}{3!}(3x)(3x)(3x) + \cdots$

$= 1 + (-\tfrac{2}{3})(3x) + \dfrac{(-\frac{2}{3})(-\frac{5}{3})}{2!}(9x^2) + \dfrac{(-\frac{2}{3})(-\frac{5}{3})(-\frac{8}{3})}{3!}(27x^3) + \cdots$

$= 1 - 2x + 5x^2 - \tfrac{40}{3}x^3 + \cdots$

Note: The coefficients are $1, -2, 5, -\frac{40}{3}, \ldots$

Step 4: State when the expansion is valid.

The expansion is valid when $|3x| < 1$, i.e. when $|x| < \tfrac{1}{3}$.

Tip: $-1 < 3x < 1$, so $-\tfrac{1}{3} < x < \tfrac{1}{3}$

Approximations

The binomial expansion is often used to make approximations.

Example 3.2

a Find the binomial expansion of $\sqrt{1-4x}$ in ascending powers of x up to and including the term in x^3.

b By substituting $x = 0.02$ in your expansion, find an approximation to $\sqrt{23}$, giving your answer to five significant figures.

Step 1: Write in index form.

a $\sqrt{1-4x} = (1-4x)^{\frac{1}{2}}$

Tip: Replace n by $\frac{1}{2}$ and x by $-4x$ in the expansion.

Step 2: Substitute into the binomial expansion.

$= 1 + (\tfrac{1}{2})(-4x) + \dfrac{(\frac{1}{2})(-\frac{1}{2})}{2!}(-4x)^2 + \dfrac{(\frac{1}{2})(-\frac{1}{2})(-\frac{3}{2})}{3!}(-4x)^3 + \cdots$

Tip: If you need to you can continue to include the extra line of working showing $(-4x)^2 = (-4x)(-4x)$ and $(-4x)^3 = (-4x)(-4x)(-4x)$.

Step 3: Simplify the terms.

$= 1 + (\tfrac{1}{2})(-4x) + \dfrac{(\frac{1}{2})(-\frac{1}{2})}{2!}(16x^2) + \dfrac{(\frac{1}{2})(-\frac{1}{2})(-\frac{3}{2})}{3!}(-64x^3) + \cdots$

$= 1 - 2x - 2x^2 - 4x^3 + \cdots$

Step 4: State when the expansion is valid.

The expansion is valid when $|-4x| < 1$, i.e. when $|x| < \tfrac{1}{4}$.

Step 1: Substitute the given value into $\sqrt{1-4x}$.

b When $x = 0.02$,

$\sqrt{1 - 4 \times 0.02} = \sqrt{0.92}$

Step 2: Simplify the surds.

$= \sqrt{\dfrac{92}{100}}$

$= \sqrt{\dfrac{4 \times 23}{100}}$

$= \dfrac{2\sqrt{23}}{10}$

Tip: Rewrite 0.92 as $\dfrac{92}{100}$.

Recall: Manipulation of surds (C1 Section 1.1).

Step 3: Substitute the given value into your expansion.

$\sqrt{1 - 4 \times 0.02} \approx 1 - 2 \times 0.02 - 2(0.02)^2 - 4(0.02)^3$

$\approx 1 - 0.04 - 0.0008 - 0.000\,032$

$\approx 0.959\,168$

Note: $x = 0.02$ is valid since $0.02 < \tfrac{1}{4}$.

Step 4: Equate the values and rearrange to find the required value.

So $\dfrac{2\sqrt{23}}{10} \approx 0.959\,168$

$\sqrt{23} \approx \dfrac{0.959\,168 \times 10}{2}$

$\sqrt{23} \approx 4.795\,84 \approx 4.7958$ (5 s.f.)

Tip: You must show your method but you can check your answer by finding $\sqrt{23}$ on your calculator.

Expanding $(a + bx)^n$

The binomial expansion of $(1 + x)^n$ can be used to expand expressions of the form $(a + bx)^n$. You must take out a factor of a^n so that the expression to be expanded is in the required format.

Tip:
$(a + bx)^n = \left[a\left(1 + \frac{b}{a}x\right)\right]^n$
$= a^n\left(1 + \frac{b}{a}x\right)^n$

Example 3.3 For each of the following expressions find the binomial expansion in ascending powers of x up to and including the term in x^3. State the range of values of x for which each expansion is valid.

a $(8 + 3x)^{\frac{1}{3}}$ **b** $\dfrac{1-x}{(2+x)^2}$

Tip: A common mistake is to forget that 8 should be to the power $\frac{1}{3}$ too.

Step 1: Take out a factor. **a** $(8 + 3x)^{\frac{1}{3}} = [8(1 + \frac{3}{8}x)]^{\frac{1}{3}} = 8^{\frac{1}{3}}(1 + \frac{3}{8}x)^{\frac{1}{3}}$

Step 2: Substitute into the binomial expansion.
$= 8^{\frac{1}{3}}\left\{1 + (\frac{1}{3})(\frac{3}{8}x) + \frac{(\frac{1}{3})(-\frac{2}{3})}{2!}(\frac{3}{8}x)^2 + \frac{(\frac{1}{3})(-\frac{2}{3})(-\frac{5}{3})}{3!}(\frac{3}{8}x)^3 + \cdots\right\}$

Tip: Replace n by $\frac{1}{3}$ and x by $\frac{3}{8}x$.

Note: $8^{\frac{1}{3}} = \sqrt[3]{8} = 2$

Step 3: Simplify the terms.
$= 2\left\{1 + (\frac{1}{3})(\frac{3}{8}x) + \frac{(\frac{1}{3})(-\frac{2}{3})}{2!}(\frac{9}{64}x^2) + \frac{(\frac{1}{3})(-\frac{2}{3})(-\frac{5}{3})}{3!}(\frac{27}{512}x^3) + \cdots\right\}$

$= 2\{1 + \frac{1}{8}x - \frac{1}{64}x^2 + \frac{5}{1536}x^3 + \cdots\}$

Step 4: Multiply through by the factor.
$= 2 + \frac{1}{4}x - \frac{1}{32}x^2 + \frac{5}{768}x^3 + \cdots$

Tip: Don't forget to multiply all the terms by 2.

Step 5: State when the expansion is valid.
The expansion is valid when $|\frac{3}{8}x| < 1$, i.e. when $|x| < \frac{8}{3}$.

Tip: Make sure you use $\frac{3}{8}x$ and not $3x$ here.

Step 1: Write the denominator in index form. **b** $\dfrac{1-x}{(2+x)^2} = (1-x)(2+x)^{-2}$

Step 2: Take out a factor.
$= (1-x)[2(1 + \frac{1}{2}x)]^{-2}$

Tip: A common mistake is to forget that 2 should be to the power -2 as well.

$= (1-x)(2^{-2})(1 + \frac{1}{2}x)^{-2}$

Note: $2^{-2} = \dfrac{1}{2^2} = \dfrac{1}{4}$

Step 3: Substitute into the binomial expansion.
$= \frac{1}{4}(1-x)\left\{1 + (-2)(\frac{1}{2}x) + \frac{(-2)(-3)}{2!}(\frac{1}{2}x)^2\right.$

$\left. + \frac{(-2)(-3)(-4)}{3!}(\frac{1}{2}x)^3 + \cdots\right\}$

Tip: Replace n by -2 and x by $\frac{1}{2}x$.

Step 4: Simplify the terms.
$= \frac{1}{4}(1-x)\left\{1 + (-2)(\frac{1}{2}x) + \frac{(-2)(-3)}{2!}(\frac{1}{4}x^2)\right.$

$\left. + \frac{(-2)(-3)(-4)}{3!}(\frac{1}{8}x^3) + \cdots\right\}$

Step 5: Expand the brackets.
$= \frac{1}{4}(1-x)\{1 - x + \frac{3}{4}x^2 - \frac{1}{2}x^3 + \cdots\}$

$= \frac{1}{4}\{1 - x + \frac{3}{4}x^2 - \frac{1}{2}x^3 - x + x^2 - \frac{3}{4}x^3 + \cdots\}$

Tip: You are only asked for four terms so you can ignore the last one when you expand the brackets.

Step 6: Collect terms.
$= \frac{1}{4}\{1 - 2x + \frac{7}{4}x^2 - \frac{5}{4}x^3 + \cdots\}$

Step 7: Multiply through by the factor.
$= \frac{1}{4} - \frac{1}{2}x + \frac{7}{16}x^2 - \frac{5}{16}x^3 + \cdots$

Tip: It's easier to expand the brackets and then multiply through by $\frac{1}{4}$.

Step 8: State when the expansion is valid.
The expansion is valid when $|\frac{1}{2}x| < 1$, i.e. when $|x| < 2$.

Using partial fractions

Expressions that can be written in partial fractions can be expanded one part at a time, as in the following example.

Example 3.4
a Express $\dfrac{1}{(1-x)(1+2x)}$ in partial fractions.

b Hence find the first four terms of the binomial expansion of $\dfrac{1}{(1-x)(1+2x)}$ in ascending powers of x.

c State the range of values of x for which the expansion is valid.

Note: To find the binomial expansion in part **b**, the given fraction could be written as $(1-x)^{-1}(1+2x)^{-1}$ and the product of two separate expansions found. The use of the word 'hence', however, means that you must use the partial fractions found in part **a**.

Step 1: Set out the partial fractions using the factors of the denominator to make new denominators.

a Let $\dfrac{1}{(1-x)(1+2x)} \equiv \dfrac{A}{1-x} + \dfrac{B}{1+2x}$

Recall: Partial fractions (Section 1.3).

Step 2: Add the fractions.

$\dfrac{1}{(1-x)(1+2x)} \equiv \dfrac{A(1+2x) + B(1-x)}{(1-x)(1+2x)}$

Step 3: Equate the numerators.

So $1 \equiv A(1+2x) + B(1-x)$

Step 4: To find A, substitute a value of x that will make the coefficient of B zero.

Substituting $x = 1$, $\quad 1 = A(1 + 2 \times 1) + B \times 0$
$1 = 3A$
$A = \tfrac{1}{3}$

Step 5: To find B, substitute a value of x that will make the coefficient of A zero.

Substituting $x = -\tfrac{1}{2}$, $\quad 1 = A \times 0 + B(1 - (-\tfrac{1}{2}))$
$1 = \tfrac{3}{2} B$
$B = \tfrac{2}{3}$

Step 6: Write out the partial fractions.

Therefore $\dfrac{1}{(1-x)(1+2x)} \equiv \dfrac{1}{3(1-x)} + \dfrac{2}{3(1+2x)}$

Step 1: Write the denominators in index form.

b $\dfrac{1}{(1-x)(1+2x)} \equiv \dfrac{1}{3(1-x)} + \dfrac{2}{3(1+2x)}$
$\equiv \tfrac{1}{3}(1-x)^{-1} + \tfrac{2}{3}(1+2x)^{-1}$

Tip: Use your answer from part **a**.

Step 2: Expand the first fraction by substituting into the binomial expansion.

$\tfrac{1}{3}(1-x)^{-1} = \tfrac{1}{3}\left\{ 1 + (-1)(-x) + \dfrac{(-1)(-2)}{2!}(-x)^2 + \dfrac{(-1)(-2)(-3)}{3!}(-x)^3 + \cdots \right\}$

Tip: Replace n by -1 and x by $-x$.

Step 3: Simplify the terms.

$= \tfrac{1}{3}\{1 + x + x^2 + x^3 + \cdots\}$
$= \tfrac{1}{3} + \tfrac{1}{3}x + \tfrac{1}{3}x^2 + \tfrac{1}{3}x^3 + \cdots$

Step 4: Expand the second fraction by substituting into the binomial expansion.

$\tfrac{2}{3}(1+2x)^{-1} = \tfrac{2}{3}\left\{ 1 + (-1)(2x) + \dfrac{(-1)(-2)}{2!}(2x)^2 + \dfrac{(-1)(-2)(-3)}{3!}(2x)^3 + \cdots \right\}$

Tip: Replace n by -1 and x by $2x$.

Step 5: Simplify the terms.

$= \tfrac{2}{3}\{1 - 2x + 4x^2 - 8x^3 + \cdots\}$
$= \tfrac{2}{3} - \tfrac{4}{3}x + \tfrac{8}{3}x^2 - \tfrac{16}{3}x^3 + \cdots$

Step 6: Combine the expansions.

$\dfrac{1}{(1-x)(1+2x)} = \tfrac{1}{3}(1-x)^{-1} + \tfrac{2}{3}(1+2x)^{-1}$
$= \tfrac{1}{3} + \tfrac{1}{3}x + \tfrac{1}{3}x^2 + \tfrac{1}{3}x^3 + \tfrac{2}{3} - \tfrac{4}{3}x + \tfrac{8}{3}x^2 - \tfrac{16}{3}x^3 + \cdots$
$= 1 - x + 3x^2 - 5x^3 + \cdots$

Tip: You could try checking your answer by expanding $(1-x)^{-1}(1+2x)^{-1}$.

Step 1: State when each expansion is valid.

c The expansion of $\frac{1}{3}(1-x)^{-1}$ is valid when $|-x| < 1$, i.e. $|x| < 1$.

The expansion of $\frac{2}{3}(1+2x)^{-1}$ is valid when $|2x| < 1$, i.e. $|x| < \frac{1}{2}$.

Step 2: Decide which values of x satisfy both inequalities.

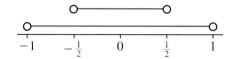

TIP:
Use an open circle to show that x cannot equal the endpoints of the set of values.

Both expansions are valid when $|x| < \frac{1}{2}$.

SKILLS CHECK 3A: Binomial series

1 For each of the following expressions find the binomial expansion in ascending powers of x up to and including the term in x^3. State the range of values of x for which each expansion is valid.

a $\dfrac{1}{(1-x)^2}$ **b** $\sqrt[3]{1-9x}$ **c** $\dfrac{1}{(1-4x)^{\frac{3}{2}}}$

 2 For each of the following expressions find the binomial expansion in ascending powers of x up to and including the term in x^2. State the ranges of values of x for which each expansion is valid.

a $\dfrac{1}{(2x-1)^3}$ **b** $(4+x)^{-\frac{3}{2}}$

3 Find the first four terms in the expansion of $\dfrac{2+x}{(1+x)^2}$ in ascending powers of x.

 4 Find the term in x^2 in the expansion of $\sqrt{\dfrac{1+x}{1-2x}}$.

5 Find the binomial expansion of $\sqrt{1+x}$ in ascending powers of x up to and including the term in x^3. By substituting $x = 0.08$ into your expansion, find an approximation to $\sqrt{3}$, giving your answer to five significant figures.

6 a Express $\dfrac{5x+4}{(1+2x)(1-x)}$ in the form $\dfrac{A}{1+2x} + \dfrac{B}{1-x}$, where A and B are constants to be found.

b Hence find the first three terms, in ascending powers of x, of the binomial expansion of $\dfrac{5x+4}{(1+2x)(1-x)}$.

c State the range of values of x for which the expansion is valid.

7 a Express $\dfrac{2x^2-5x}{(2+x)(1-x)^2}$ in the form $\dfrac{A}{2+x} + \dfrac{B}{1-x} + \dfrac{C}{(1-x)^2}$, where A, B and C are constants to be found.

b Hence find the first three terms of the binomial expansion of $\dfrac{2x^2-5x}{(2+x)(1-x)^2}$ in ascending powers of x.

c State the range of values of x for which the expansion is valid.

8 In the expansion of $(1 + ax)^n$, in ascending powers of x, the coefficient of x is 2 and the coefficient of x^3 is twice the coefficient of x^2.

   a Find a and n.

   b Find the coefficient of x^3.

   c State the values of x for which the expansion is valid.

SKILLS CHECK **3A EXTRA** is on the CD

Examination practice 3: Sequences and series

1 a Obtain the binomial expansion in ascending powers of x up to and including the term in x^3 of the following, giving each term in its simplest form.

   i $(1 + x)^{-1}$ **ii** $(1 + 4x)^{\frac{1}{2}}$

   b Hence show that, for small values of x,

   $$2(1 + 4x)^{\frac{1}{2}} + \frac{4}{1 + x} \approx 6 + kx^3$$

   where k is a constant to be found. [AQA June 2004]

2 a Find the binomial expansion of $(1 + \frac{1}{2}x)^{-2}$ in ascending powers of x up to and including the term in x^3.

   b Hence determine the coefficient of x^3 in the expansion of $(2 + 3x)(1 + \frac{1}{2}x)^{-2}$.

3 a Use the binomial series to expand $(3 + 2x)^{-3}$, in ascending powers of x, up to and including the term in x^3. Give each coefficient as a simplified fraction.

   b State the values of x for which the expansion is valid.

4 a Express $\dfrac{4 - x}{(1 - x)(2 + x)}$ in the form $\dfrac{A}{1 - x} + \dfrac{B}{2 + x}$.

   b i Show that the first **three** terms in the expansion of

   $$\frac{1}{2 + x}$$

   in ascending powers of x are $\dfrac{1}{2} - \dfrac{x}{4} + \dfrac{x^2}{8}$.

   ii Obtain also the first **three** terms in the expansion of

   $$\frac{1}{1 - x}$$

   in ascending powers of x.

   c Hence, or otherwise, obtain the first **three** terms in the expansion of

   $$\frac{4 - x}{(1 - x)(2 + x)}$$

   in ascending powers of x. [AQA June 2002]

5 a i Obtain the first **four** terms in the binomial expansion of $\dfrac{1}{1+x}$ in ascending powers of x.

  ii Show that the first four terms in the binomial expansion of $\dfrac{1}{3+2x}$ in ascending powers of x are
$$\dfrac{1}{3} - \dfrac{2}{9}x + \dfrac{4}{27}x^2 - \dfrac{8}{81}x^3.$$

b Express $\dfrac{8+7x}{(1+x)(3+2x)}$ in the form $\dfrac{A}{1+x} + \dfrac{B}{3+2x}$.

c Hence obtain the first **four** terms in the expansion of $\dfrac{8+7x}{(1+x)(3+2x)}$ in ascending powers of x.

[AQA Jan 2005]

6 It is given that
$$\dfrac{(2x-3)}{(1-x)(2-x)} \equiv \dfrac{A}{1-x} + \dfrac{B}{2-x}.$$

a Find the values of the constants A and B.

b Hence, or otherwise, find the series expansion in ascending powers of x, up to and including the term in x^3, of $\dfrac{(2x-3)}{(1-x)(2-x)}$.

c State the values of x for which the expansion is valid.

7 When $(1+kx)^n$ is expanded as a series in ascending powers of x, the first three terms are $1 - x + \tfrac{3}{2}x^2$.

a Find the value of k and the value of n.

b Find the coefficient of x^3.

c State the set of values of x for which the expansion is valid.

4 Trigonometry

4.1 Addition formulae

Use of the formulae for $\sin(A \pm B)$, $\cos(A \pm B)$ and $\tan(A \pm B)$.

For angles A and B, the **addition formulae** give expressions for sin, cos and tan of the **sum** or **difference** of A and B. They are identities and so they are true for all values of A and B.

The six formulae are summarised in your formulae booklet as follows:

$$\sin(A \pm B) = \sin A \cos B \pm \cos A \sin B$$

$$\cos(A \pm B) = \cos A \cos B \mp \sin A \sin B$$

$$\tan(A \pm B) = \frac{\tan A \pm \tan B}{1 \mp \tan A \tan B}$$

Note: These formulae relate to two angles added or subtracted, not just added. They are also called the **compound angle formulae**.

Tip: Make sure that you understand how to use the $\pm$ and $\mp$ symbols, for example $\cos(A - B) \equiv \cos A \cos B + \sin A \sin B$.

Example 4.1

a If $\sin \theta = \cos(\theta + 30°)$, show that $\tan \theta = \frac{1}{3}\sqrt{3}$, given that $\cos 30° = \frac{1}{2}\sqrt{3}$ and $\sin 30° = \frac{1}{2}$.

b Hence find all the values of θ, where $0° < \theta < 360°$, for which $\sin \theta = \cos(\theta + 30°)$.

Step 1: Expand the sum using the addition formula.
Step 2: Insert known trig values.
Step 3: Simplify to the required form.

a
$$\sin \theta = \cos(\theta + 30°)$$
$$\Rightarrow \sin \theta = \cos \theta \cos 30° - \sin \theta \sin 30°$$
$$\Rightarrow \sin \theta = \cos \theta \times \tfrac{1}{2}\sqrt{3} - \sin \theta \times \tfrac{1}{2}$$
$$(\times 2) \quad 2 \sin \theta = \sqrt{3} \cos \theta - \sin \theta$$
$$3 \sin \theta = \sqrt{3} \cos \theta$$
$$(\div \text{ by } 3 \cos \theta)$$
$$\frac{3 \sin \theta}{3 \cos \theta} = \frac{\sqrt{3} \cos \theta}{3 \cos \theta}$$
$$\tan \theta = \tfrac{1}{3}\sqrt{3}$$

Note: It is permissible to divide by $\cos \theta$ here, since $\cos \theta = 0$ is clearly not a solution of the equation.

Recall: $\frac{\sin \theta}{\cos \theta} \equiv \tan \theta$ (C2 Section 3.6).

Step 1: Use the result from part **a** to solve the simple trig equation.

b
$$\sin \theta = \cos(\theta + 30°)$$
$$\Rightarrow \tan \theta = \tfrac{1}{3}\sqrt{3}$$
$$\Rightarrow \theta = 30°, 210°$$

Tip: The values in range are PV and PV + 180°, since the tan function repeats every 180° (C2 Section 3.6).

Example 4.2 Prove the identity

$$\cos A \cos(A - B) + \sin A \sin(A - B) \equiv \cos B$$

Step 1: Expand the LHS using the addition formulae.

$$\text{LHS} = \cos A(\cos A \cos B + \sin A \sin B)$$
$$+ \sin A(\sin A \cos B - \cos A \sin B)$$
$$= \cos^2 A \cos B + \cos A \sin A \sin B$$
$$+ \sin^2 A \cos B - \sin A \cos A \sin B$$

Step 2: Simplify and use appropriate identities to arrive at the RHS.

$$= \cos^2 A \cos B + \sin^2 A \cos B$$
$$= \cos B(\cos^2 A + \sin^2 A)$$
$$= \cos B$$
$$= \text{RHS}$$

So $\cos A \cos(A - B) + \sin A \sin(A - B) \equiv \cos B$.

Tip: You will not gain any marks if all you do is quote the formulae for $\cos(A - B)$ and $\sin(A - B)$ given in the formulae booklet. You must then go on to use them in the question.

Tip: $\cos^2 A + \sin^2 A \equiv 1$ (C2 Section 3.6).

Tip: Keep going, even when the expression looks very complicated. Often terms cancel or you can use other identities to simplify further.

25

Example 4.3 For obtuse angle A and acute angle B, it is given that $\sin A = \frac{2}{3}$ and $\cos B = \frac{3}{5}$. Find the exact values of the following, simplifying your answers:

    a $\cos A$ **b** $\sin B$

    c $\cos(A - B)$ **d** $\sin(A + B)$

Step 1: Use a trig identity connecting $\sin \theta$ and $\cos \theta$.

a $\sin^2 A + \cos^2 A = 1$
$$\Rightarrow \cos^2 A = 1 - (\tfrac{2}{3})^2 = \tfrac{5}{9}$$
Since A is obtuse, $\cos A < 0$, so $\cos A = -\sqrt{\tfrac{5}{9}} = -\tfrac{\sqrt{5}}{3}$.

Tip: You must not calculate the angles. Use trig identities and leave your answer in surd form where necessary.

Step 1: Use a trig identity connecting $\sin \theta$ and $\cos \theta$.

b $\sin^2 B + \cos^2 B = 1$
$$\Rightarrow \sin^2 B = 1 - (\tfrac{3}{5})^2 = \tfrac{16}{25}$$
Since B is acute, $\sin B > 0$, so $\sin B = \sqrt{\tfrac{16}{25}} = \tfrac{4}{5}$.

Step 1: Expand using an appropriate addition formula.

Step 2: Simplify.

c $\cos(A - B) = \cos A \cos B + \sin A \sin B$
$$= -\tfrac{\sqrt{5}}{3} \times \tfrac{3}{5} + \tfrac{2}{3} \times \tfrac{4}{5}$$
$$= -\tfrac{3\sqrt{5}}{15} + \tfrac{8}{15}$$
$$= \tfrac{1}{15}(8 - 3\sqrt{5})$$

Step 1: Expand using an appropriate addition formula.

Step 2: Simplify.

d $\sin(A + B) = \sin A \cos B + \cos A \sin B$
$$= \tfrac{2}{3} \times \tfrac{3}{5} + \left(-\tfrac{\sqrt{5}}{3}\right) \times \tfrac{4}{5}$$
$$= \tfrac{6}{15} - \tfrac{4\sqrt{5}}{15}$$
$$= \tfrac{1}{15}(6 - 4\sqrt{5})$$

Example 4.4 Use the expansions of $\sin(A - B)$ and $\cos(A - B)$ to derive the formula for $\tan(A - B)$ in terms of $\tan A$ and $\tan B$.

Step 1: Use an appropriate trig identity connecting $\sin(A - B)$ and $\cos(A - B)$.

Step 2: Expand using the double angle formulae.

Step 3: Divide the numerator and denominator by an appropriate term.

$$\tan(A - B) = \frac{\sin(A - B)}{\cos(A - B)}$$

$$= \frac{\sin A \cos B - \cos A \sin B}{\cos A \cos B + \sin A \sin B}$$

$$= \frac{\dfrac{\sin A \cos B}{\cos A \cos B} - \dfrac{\cos A \sin B}{\cos A \cos B}}{\dfrac{\cos A \cos B}{\cos A \cos B} + \dfrac{\sin A \sin B}{\cos A \cos B}}$$

$$= \frac{\tan A - \tan B}{1 + \tan A \tan B}$$

So $\tan(A - B) = \dfrac{\tan A - \tan B}{1 + \tan A \tan B}$.

Tip: Dividing by $\cos A \cos B$ gives the required format.

4.2 Expressions for $a \cos \theta + b \sin \theta$

Use of expressions for $a \cos \theta + b \sin \theta$ in the equivalent forms of $r \cos(\theta \pm \alpha)$ or $r \sin(\theta \pm \alpha)$.

The expression $a \cos \theta + b \sin \theta$ can be written in the form $r \cos(\theta \pm \alpha)$ or $r \sin(\theta \pm \alpha)$.

Note: This form is useful for solving equations, drawing graphs and finding maximum and minimum values of functions.

Example 4.5 It is given that $f(\theta) = 2\cos\theta + 3\sin\theta$.

a Express $f(\theta)$ in the form $r\cos(\theta - \alpha)$, where $r > 0$ and $0° < \alpha < 90°$.

b Hence solve the equation $2\cos\theta + 3\sin\theta = 3$, where $0° < \theta < 360°$. If an answer is not exact, give it correct to one decimal place.

Step 1: Use an appropriate identity to expand the trig expression.

a
$$r\cos(\theta - \alpha) \equiv 2\cos\theta + 3\sin\theta$$
$$\Rightarrow r(\cos\theta\cos\alpha + \sin\theta\sin\alpha) \equiv 2\cos\theta + 3\sin\theta$$
$$r\cos\theta\cos\alpha + r\sin\theta\sin\alpha \equiv 2\cos\theta + 3\sin\theta$$

Recall:
$\cos(A - B)$
$\equiv \cos A\cos B + \sin A\sin B$
(Section 4.1).

Step 2: Equate coefficients to form two equations that enable you to work out r and α.

Equating coefficients of $\cos\theta$:
$$r\cos\alpha = 2 \quad \text{①}$$
Equating coefficients of $\sin\theta$:
$$r\sin\alpha = 3 \quad \text{②}$$

Step 3: Divide the equations to find α.

Equation ② ÷ equation ① gives
$$\frac{r\sin\alpha}{r\cos\alpha} = \frac{3}{2}$$
$$\Rightarrow \tan\alpha = \frac{3}{2}$$
$$\alpha = 56.30\ldots°$$

Recall:
$\frac{\sin\alpha}{\cos\alpha} \equiv \tan\alpha$

Tip:
$\alpha = \tan^{-1}\left(\frac{b}{a}\right) = \tan^{-1}\left(\frac{3}{2}\right)$

Step 4: Square and add the equations to find r.

Squaring ① and ② and adding gives
$$r^2\cos^2\alpha + r^2\sin^2\alpha = 2^2 + 3^2$$
$$r^2(\cos^2\alpha + \sin^2\alpha) = 13$$
$$r^2 = 13$$
$$r = \sqrt{13}$$

Recall:
$\cos^2\alpha + \sin^2\alpha \equiv 1$

Tip:
$r = \sqrt{a^2 + b^2} = \sqrt{2^2 + 3^2}$

Step 5: Write $f(\theta)$ in the required format.

So $f(\theta) = 2\cos\theta + 3\sin\theta \equiv \sqrt{13}\cos(\theta - 56.30\ldots°)$

Step 6: Use the format found in **a** to form a simple trig equation and solve.

b
$$2\cos\theta + 3\sin\theta = 3$$
$$\Rightarrow \sqrt{13}\cos(\theta - 56.30\ldots°) = 3$$
$$\cos(\theta - 56.30\ldots°) = \frac{3}{\sqrt{13}} = 0.8320\ldots$$

Let $x = \theta - 56.30\ldots°$.
The equation becomes $\cos x = 0.8320\ldots$
Now $0° < \theta < 360°$
so $-56.30° < x < 303.69°$
PV of $x = 33.69\ldots°$

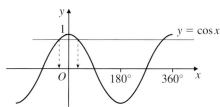

Tip:
Consider the range required for x. In this case you will need to consider negative values.

Note:
The next positive value is $360° - 33.69\ldots° = 326.30\ldots°$ which is out of range.

The other value in range is $-33.69\ldots°$.
So $\theta - 56.30\ldots° = -33.69\ldots°, 33.69\ldots°$
$$\theta = 22.6° \text{ (1 d.p.)}, 90°$$

Note that if you are given that $0 < \alpha < \frac{1}{2}\pi$ then you must work in **radians**.

In part **a** of Example 4.5, $\alpha = 0.983^c$ (3 d.p.)

so $f(\theta) = 2\cos\theta + 3\sin\theta \equiv \sqrt{13}\cos(\theta - 0.983^c)$

In part **b**, for values of θ in the interval

$$0 < \theta < 2\pi$$

letting $x = \theta - 0.983^c$ gives

$$-0.983^c < x < 5.300^c$$

For the equation $\cos x = 0.8320\ldots$, the PV of $x = 0.588\ldots^c$ and the other value of x in range is $-0.588\ldots^c$

So $\theta - 0.983^c = -0.588\ldots^c, 0.588\ldots^c$

$\Rightarrow \quad \theta = 0.39^c$ (2 d.p.), 1.57^c (2 d.p.)

> *Tip:*
> Make sure that you are confident using radians.

Example 4.6 It is given that $f(x) = 3\sin x - 3\cos x$.

a Find the values of r and α such that $f(x) \equiv r\sin(x - \alpha)$, where $r > 0$ and $0° < \alpha < 90°$. Give the value of r exactly.

b The diagram shows a sketch of $y = f(x)$ for $0° \leq x \leq 360°$. The curve crosses the y-axis at A and the x-axis at B and C. There is a maximum point at D and a minimum point at E.

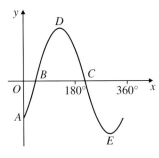

i Describe a sequence of geometric transformations that map the curve $y = \sin x$ onto the curve $y = f(x)$.
ii Find the coordinates of A, B and C.
iii Find the coordinates of the maximum point D and the minimum point E.

Step 1: Use an appropriate identity to expand the trig expression.

a
$$r\sin(x - \alpha) \equiv 3\sin x - 3\cos x$$
$$\Rightarrow r(\sin x \cos\alpha - \cos x \sin\alpha) \equiv 3\sin x - 3\cos x$$
$$r\sin x \cos\alpha - r\cos x \sin\alpha \equiv 3\sin x - 3\cos x$$

> *Recall:*
> $\sin(A - B)$
> $\equiv \sin A \cos B - \cos A \sin B$
> (Section 4.1)

Step 2: Equate coefficients to form two equations that enable you to work out r and α.

Equating coefficients of $\sin x$:
$$r\cos\alpha = 3 \qquad \qquad ①$$
Equating coefficients of $\cos x$:
$$r\sin\alpha = 3 \qquad \qquad ②$$

Equation ② ÷ equation ① gives

$$\frac{r\sin\alpha}{r\cos\alpha} = \frac{3}{3}$$

$\Rightarrow \quad \tan\alpha = 1$

$\alpha = 45°$

> *Recall:*
> $\dfrac{\sin\alpha}{\cos\alpha} = \tan\alpha$

Squaring ① and ② and adding gives
$$r^2 \cos^2 \alpha + r^2 \sin^2 \alpha = 3^2 + 3^2$$
$$r^2 (\cos^2 \alpha + \sin^2 \alpha) = 18$$
$$r^2 = 18$$
$$r = \sqrt{18} = \sqrt{9 \times 2} = 3\sqrt{2}$$

So $f(x) = 3\sqrt{2} \sin(x - 45°)$, with $r = 3\sqrt{2}$ and $\alpha = 45°$.

Recall:
$\cos^2 \alpha + \sin^2 \alpha = 1$

Tip:
Leave your answer in surd form. It is good practice to simplify it.

Step 1: Describe appropriate transformations for $y = af(x - b)$.

b i To transform the curve $y = \sin x$ to $y = 3\sqrt{2} \sin(x - 45°)$:
- translate by $45°$ in the positive x-direction
- stretch by scale factor $3\sqrt{2}$ in the y-direction.

Recall:
Transformations
(C3 Section 1.5).

Step 2: Identify the coordinates of the intercepts with the axes by setting $x = 0$ and $y = 0$.

ii $y = 3\sqrt{2} \sin(x - 45°)$
When $x = 0$, $y = 3\sqrt{2} \times \sin(-45°) = -3$
So the coordinates of A are $(0, -3)$.

When $y = 0$, $3\sqrt{2} \sin(x - 45°) = 0$
$\Rightarrow \quad x - 45° = 0 \quad$ or $\quad x - 45° = 180°$
$\Rightarrow \quad\quad x = 45° \quad\quad\quad x = 225°$

So B is the point $(45°, 0)$ and C is the point $(225°, 0)$.

Recall:
$\sin 0° = \sin 180° = 0$

Step 3: Identify the coordinates of the maximum and minimum points by considering the range of the transformed curve.

iii The maximum value of $\sin x$ is 1 and it occurs when $x = 90°$.
Hence the maximum value of $3\sqrt{2} \sin(x - 45°)$ is $3\sqrt{2} \times 1 = 3\sqrt{2}$.
The maximum value occurs when $x - 45° = 90° \Rightarrow x = 135°$.
So D is the point $(135°, 3\sqrt{2})$.

Recall:
$\sin 90° = 1$

The minimum value of $\sin x$ is -1 and it occurs when $x = 270°$.
Hence the minimum value of $3\sqrt{2} \sin(x - 45°)$ is $3\sqrt{2} \times (-1) = -3\sqrt{2}$.
The minimum value occurs when $x - 45° = 270° \Rightarrow x = 315°$.
So E is the point $(315°, -3\sqrt{2})$.

Recall:
$\sin 270° = -1$

Which alternative format is preferable?
If the alternative format of $a \cos \theta + b \sin \theta$ is not specified in the question, it is advisable to ensure that α is acute, that is $0° < \alpha < 90°$ or $0 < \alpha < \frac{1}{2}\pi$.

For example,
$$3 \cos \theta - 4 \sin \theta \equiv r \cos(\theta + \alpha)$$
and
$$4 \sin \theta - 3 \cos \theta \equiv r \sin(\theta - \alpha)$$

However, for
$$3 \cos \theta + 4 \sin \theta$$

you could use
$$3 \cos \theta + 4 \sin \theta \equiv r \cos(\theta - \alpha)$$

or
$$4 \sin \theta + 3 \cos \theta \equiv r \sin(\theta + \alpha)$$

Tip:
In the cos format, the sign in the middle is opposite to the sign in the expression. In the sin format, the sign in the middle is the same as the sign in the expression.

SKILLS CHECK 4A: Addition formulae and expressions for $a\cos\theta + b\sin\theta$

1. Given that angles A and B are acute and that $\sin A = \frac{4}{5}$ and $\cos B = \frac{12}{13}$, find the exact values of:
   - **a** $\cos A$
   - **b** $\sin B$
   - **c** $\cos(A - B)$
   - **d** $\sec(A - B)$

2. **a** Given that $\sin\theta = \dfrac{1}{\sqrt{2}}$ and θ is acute, find the exact value of $\cos\theta$.

   b Show that $\sin(x + 45°) = \dfrac{1}{\sqrt{2}}(\sin x + \cos x)$.

   c Hence solve the equation $\sin(x + 45°) = \sqrt{2}\cos x$, for values of x in the interval $0° \leq x \leq 360°$.

3. **a** Simplify
   - **i** $\sin(A + B) + \sin(A - B)$
   - **ii** $\cos(A + B) + \cos(A - B)$

   b Hence prove the identity $\dfrac{\sin(A + B) + \sin(A - B)}{\cos(A + B) + \cos(A - B)} \equiv \tan A$.

4. **a** Expand $\sin(X - Y)$.

   b By letting $X = 4A$ and $Y = 2A$, or otherwise, prove the identity
   $$\frac{\sin 4A \cos 2A - \cos 4A \sin 2A}{\sin A} \equiv 2\cos A.$$

5. Given that $\tan(A + B) = 1$ and $\tan A = \frac{1}{3}$, find the value of $\tan B$.

6. **a** Prove by a counter-example that the statement '$\cot(A + B) \equiv \cot A + \cot B$ for all A and B' is false.

   b Prove that $\cot(A + B) \equiv \dfrac{\cot A \cot B - 1}{\cot A + \cot B}$.

7. **a** Express $2\cos x + \sin x$ in the form $r\cos(x - \alpha)$, where $r > 0$ and $0° < \alpha < 90°$.

   b Hence solve the equation $2\cos x + \sin x = 1$, for values of x in the interval $0° < x < 360°$.

8. $f(\theta) = 4\cos\theta - 3\sin\theta$.

   a Express $f(\theta)$ in the form $r\cos(\theta + \alpha)$, where $r > 0$ and $0 < \alpha < \frac{1}{2}\pi$, giving the value of α in radians to two decimal places.

   b Hence
   - **i** write down the maximum value of $f(\theta)$,
   - **ii** find the smallest positive value of θ at which $f(\theta)$ is maximum.

9. **a** Express $\sin x + \cos x$ in the form $r\sin(x + \alpha)$, where $r > 0$ and $0° < \alpha < 90°$.

   b Hence find all the values of x, in the interval $-180° < x < 180°$, for which
   $$\sqrt{2}\sin x + \sqrt{2}\cos x = -1$$

10. The expression $k\cos x + 15\sin x$ can be written in the form $17\cos(x - \alpha)$, where $0° < \alpha < 90°$ and $k > 0$. Find the values of α and k.

11 **a** Express $f(x) = 4\cos x + 2\sin x$ in the form $R\cos(x - \alpha)$, where R is a positive constant and $0 < \alpha < \frac{1}{2}\pi$, giving the value of R exactly and the value of α in radians to 3 decimal places.

b Hence, find the solutions in the interval $0 < \alpha < 2\pi$ of the equation
$$4\cos x + 2\sin x = 3$$
Give each solution in radians to two decimal places.

SKILLS CHECK **4A EXTRA** is on the CD

4.3 Double angle formulae

Knowledge and use of double angle formulae, including use in integration of trigonometric functions.

Putting $B = A$ in the identities for the sum of two angles gives the **double angle formulae** which are true for all values of A.

$$\sin 2A \equiv 2\sin A \cos A$$

$$\cos 2A \equiv \cos^2 A - \sin^2 A$$

$$\tan 2A \equiv \frac{2\tan A}{1 - \tan^2 A}$$

Note:
You should learn these. However, if you forget them, put $B = A$ in the formulae for $\sin(A + B)$, $\cos(A + B)$ and $\tan(A + B)$, given in the formulae booklet.

Often, alternative formats of the formula for $\cos 2A$ are used.
Using $\sin^2 A + \cos^2 A \equiv 1$

$$\cos 2A \equiv \cos^2 A - \sin^2 A$$
$$\equiv \cos^2 A - (1 - \cos^2 A)$$
$$\equiv 2\cos^2 A - 1$$

Also,
$$\cos 2A \equiv \cos^2 A - \sin^2 A$$
$$\equiv (1 - \sin^2 A) - \sin^2 A$$
$$\equiv 1 - 2\sin^2 A$$

Note:
Do not expect to always see $2A$ and A as the angles. Look out especially for *half angles*, where, for example
$$\sin A \equiv 2\sin\frac{A}{2}\cos\frac{A}{2}$$
$$\cos A \equiv 2\cos^2\frac{A}{2} - 1$$

Summarising:
$$\cos 2A \equiv \cos^2 A - \sin^2 A \equiv 2\cos^2 A - 1 \equiv 1 - 2\sin^2 A$$

Note:
You must be confident about using any of these formats.

Example 4.7 Solve the equation $\cos 2x + 3\sin x = 2$ for $0° \leq x < 360°$.

Step 1: Use an appropriate double angle formula.
$$\cos 2x + 3\sin x = 2$$
$$\Rightarrow 1 - 2\sin^2 x + 3\sin x = 2$$
$$2\sin^2 x - 3\sin x + 1 = 0$$

Tip:
Form an equation in $\sin x$ by using the version of $\cos 2x$ that involves $\sin x$.

Step 2: Solve the equation in $\sin x$.
$$(2\sin x - 1)(\sin x - 1) = 0$$
$$\Rightarrow 2\sin x - 1 = 0 \quad \text{or} \quad \sin x - 1 = 0$$
$$\sin x = 0.5 \qquad\qquad \sin x = 1$$
$$x = 30°, 150° \qquad\qquad x = 90°$$

So $x = 30°, 90°, 150°$

Tip:
PV = 30°; the other solution in range is 180° − PV.

Tip:
PV = 90°. This is the only value in range. Be careful not to include other incorrect values within the range which will lead to loss of marks.

Example 4.8 Given that $\tan x = \frac{3}{4}$, using an appropriate double angle formula find the *exact* value of $\cot 2x$.

Step 1: Expand using an appropriate double angle formula.

First find the value of $\tan 2x$.

$$\tan 2x \equiv \frac{2 \tan x}{1 - \tan^2 x}$$

Step 2: Substitute the known value.

$$= \frac{2 \times \frac{3}{4}}{1 - \left(\frac{3}{4}\right)^2} = \frac{24}{7}$$

Step 3: Use the reciprocal function.

$$\cot 2x = \frac{1}{\tan 2x} = 1 \div \frac{24}{7} = \frac{7}{24}$$

Tip: You will gain no marks for calculating $\tan^{-1}\left(\frac{3}{4}\right)$ and using the angle obtained to calculate $\cot 2x$.

Tip: If you use the fraction key on the calculator, you may need to put $\frac{3}{4}$ in a bracket before squaring.

Recall: $\cot A = \frac{1}{\tan A}$ (C3 Section 2.2).

Example 4.9 For values of x in the interval $0 \leqslant x < 2\pi$, solve the following equations, giving your answers in radians to two decimal places where appropriate.

a $3 \sin 2x = \sin x$ **b** $\sin\left(\frac{1}{2}x\right) = 3 \sin x$

Step 1: Use an appropriate double angle formula.

a
$$3 \sin 2x = \sin x$$
$$\Rightarrow \quad 6 \sin x \cos x = \sin x$$
$$6 \sin x \cos x - \sin x = 0$$

Step 2: Factorise and solve.

$$\sin x (6 \cos x - 1) = 0$$
$$\Rightarrow \quad \sin x = 0$$
$$x = 0^c, \pi \, (= 3.141\ldots^c)$$

or $\quad 6 \cos x - 1 = 0$

$$\cos x = \tfrac{1}{6}$$
$$x = 1.403\ldots^c, 4.879\ldots^c$$

So $x = 0^c, 1.40^c$ (2 d.p.), 3.14^c (2 d.p.), 4.88^c (2 d.p.)

Recall: $\sin 2A = 2 \sin A \cos A$

Tip: Factorise here. Do not divide through by $\sin x$ as this will result in the loss of some solutions.

Recall: Graph of $y = \sin x$ (C2 Section 3.5).

Recall: The values in range are PV and $2\pi - $ PV (C2 Section 3.6).

Step 1: Notice the link with part **a** and apply a suitable substitution.

b Letting $\frac{1}{2}x = \theta$, the equation

$$\sin\left(\tfrac{1}{2}x\right) = 3 \sin x, \; 0 \leqslant x < 2\pi$$

becomes

$$\sin \theta = 3 \sin 2\theta, \; 0 \leqslant \theta < \pi$$

Step 2: Solve using your answers from **a**.

From **a**, since $0 \leqslant \theta < \pi$, $\theta = 0^c$ and $1.403\ldots^c$

$$\Rightarrow \quad \tfrac{1}{2}x = 0^c, 1.403\ldots^c$$
$$\Rightarrow \quad x = 0^c, 2.806\ldots^c$$

So $x = 0^c, 2.81^c$ (2 d.p.)

Tip: Be on the look out for links between parts of questions.

Tip: Double your answers from part **a**, but only include values in the required interval.

Example 4.10 **a** By expanding $\sin(2A + A)$ prove that $\sin 3A \equiv 3\sin A - 4\sin^3 A$.

b Hence, or otherwise, solve $3\sin x - 4\sin^3 x = 1$ for values of x such that $0° < x < 180°$.

Step 1: Use the addition formula suggested.

a $\sin 3A \equiv \sin(2A + A)$
$\equiv \sin 2A \cos A + \cos 2A \sin A$

Recall: Formulae for $\sin(A + B)$, $\sin 2A$ and $\cos 2A$, $\cos^2 A$.

Step 2: Use appropriate formulae to express all terms in terms of $\sin A$.

$\equiv 2\sin A \cos A \cos A + (1 - 2\sin^2 A)\sin A$
$\equiv 2\sin A \cos^2 A + \sin A - 2\sin^3 A$
$\equiv 2\sin A(1 - \sin^2 A) + \sin A - 2\sin^3 A$
$\equiv 2\sin A - 2\sin^3 A + \sin A - 2\sin^3 A$
$\equiv 3\sin A - 4\sin^3 A$

Step 3: Use the result in **a** to form a simple equation and solve in the appropriate range.

b $3\sin x - 4\sin^3 x = 1$
$\Rightarrow \sin 3x = 1$
$\Rightarrow \quad 3x = 90°, 450°$
$x = 30°, 150°$

Range for $3x$:
$0° < x < 180°$
$\Rightarrow 0° < 3x < 540°$

Tip: Make sure that you have included all the solutions in range.

Example 4.11 Prove that $\dfrac{\sin 2A}{1 - \cos 2A} \equiv \cot A$.

Step 1: Use an appropriate double angle formula in the numerator.

$\text{LHS} = \dfrac{\sin 2A}{1 - \cos 2A}$

$= \dfrac{2\sin A \cos A}{1 - \cos 2A}$

Tip: In the numerator, use $\sin 2A = 2\sin A \cos A$.

Step 2: Use an appropriate double angle formula in the denominator.

$= \dfrac{2\sin A \cos A}{1 - (1 - 2\sin^2 A)}$

$= \dfrac{2\sin A \cos A}{2\sin^2 A}$

$= \dfrac{\cos A}{\sin A}$

$= \cot A$

$= \text{RHS}$

So $\dfrac{\sin 2A}{1 - \cos 2A} \equiv \cot A$.

Tip: In the denominator, use $\cos 2A = 1 - 2\sin^2 A$.

Tip: Do your working in stages, one thing at a time. Do not try to do too many things at once.

Example 4.12 Find the Cartesian equation of the curve defined by the parametric equations

$x = \cos\theta, \quad y = \sin 2\theta$

Step 1: Rewrite y in terms of $\sin\theta$ and $\cos\theta$.

$y = \sin 2\theta$
$= 2\sin\theta \cos\theta$

Recall: Parametric equations (Section 2.1).

Step 2: Replace $\cos\theta$ by x.

Substituting for $\cos\theta$ gives

$y = 2x\sin\theta$

Step 3: Use trig identities to connect $\sin\theta$ and x.

Since $x = \cos\theta$ and $\cos^2\theta + \sin^2\theta \equiv 1$

$$x^2 + \sin^2\theta = 1$$
$$\sin^2\theta = 1 - x^2 \quad \text{①}$$

Tip:
You need to eliminate θ, so squaring is a good method to use.

Step 4: Eliminate θ in the equation for y.

$y = 2x\sin\theta \Rightarrow y^2 = 4x^2\sin^2\theta$

Substituting $\sin^2\theta$ from ① gives

$$y^2 = 4x^2(1 - x^2)$$

The Cartesian equation of the curve is $y^2 = 4x^2(1 - x^2)$.

Note:
$y = 2x\sqrt{1-x^2}$ gives only part of the curve. The other part is given by $y = -2x\sqrt{1-x^2}$.

Use of double angle formulae when integrating trigonometric functions

In *Core 3* you integrated trigonometric functions, using the following results:

f(x)	$\int f(x)\,dx$
$\cos kx$	$\dfrac{1}{k}\sin kx + c$
$\sin kx$	$-\dfrac{1}{k}\cos kx + c$
$\sec^2 kx$	$\dfrac{1}{k}\tan kx + c$

Recall:
Integration (C3 Section 5.1).

Recall:
$\int \sec^2 kx\,dx$ is given in the formulae booklet.

The trigonometric functions that you are required to integrate are extended in *Core 4* to include ones that can be integrated using the double angle formulae.

The following rearrangements of $\cos 2A$ are useful:

$$\cos 2A \equiv 2\cos^2 A - 1 \Rightarrow \cos^2 A \equiv \tfrac{1}{2}(1 + \cos 2A)$$
$$\cos 2A \equiv 1 - 2\sin^2 A \Rightarrow \sin^2 A \equiv \tfrac{1}{2}(1 - \cos 2A)$$

Tip:
You will find it helpful to learn these.

Example 4.13 Find

a $\int \sin^2 x\,dx$ **b** $\int \cos^2 3x\,dx$

c $\int \sin 4x \cos 4x\,dx$ **d** $\int_0^{2\pi} \sin^2(\tfrac{1}{2}x)\,dx$

Step 1: Apply an appropriate trig identity.

a $\int \sin^2 x\,dx = \int \tfrac{1}{2}(1 - \cos 2x)\,dx$

$= \tfrac{1}{2}\int (1 - \cos 2x)\,dx$

Step 2: Integrate term by term.

$= \tfrac{1}{2}(x - \tfrac{1}{2}\sin 2x) + c$

Tip:
Rearrange $\cos 2A \equiv 1 - 2\sin^2 A$.

Tip:
Take out a numerical factor before integrating.

Step 1: Apply an appropriate trig identity.

b $\int \cos^2 3x\,dx = \int \tfrac{1}{2}(1 + \cos 6x)\,dx$

$= \tfrac{1}{2}\int (1 + \cos 6x)\,dx$

Step 2: Integrate term by term.

$= \tfrac{1}{2}(x + \tfrac{1}{6}\sin 6x) + c$

Tip:
Rearrange $\cos 2A \equiv 2\cos^2 A - 1$.

Step 1: Apply an appropriate trig identity.
Step 2: Integrate.

c $\int \sin 4x \cos 4x \, dx = \frac{1}{2}\int \sin 8x \, dx$
$= \frac{1}{2}(-\frac{1}{8}\cos 8x) + c$
$= -\frac{1}{16}\cos 8x + c$

Tip:
Use the double angle formula for sin 2A.

Step 1: Apply an appropriate trig identity.
Step 2: Integrate term by term.
Step 3: Substitute the limits and evaluate.

d $\int_0^{2\pi} \sin^2(\frac{1}{2}x) \, dx = \int_0^{2\pi} \frac{1}{2}(1 - \cos x) \, dx$
$= \frac{1}{2}\left[x - \sin x\right]_0^{2\pi}$
$= \frac{1}{2}(2\pi - \sin 2\pi - (0 - \sin 0))$
$= \frac{1}{2}(2\pi - 0 - (0 - 0))$
$= \pi$

Tip:
Rearrange $\cos 2A \equiv 1 - 2\sin^2 A$.

Recall:
Definite integration (C3 Section 5.1).

SKILLS CHECK 4B: Double angle formulae

1 Find the values of θ, where $0° \leq \theta \leq 360°$, such that
 a $\cos 2\theta = 1 + \sin \theta$ **b** $\sin 2\theta = \cot \theta$

2 Find the values of x, where $0 \leq x < 2\pi$, such that
 a $\cos 2x = \cos x$ **b** $\cos x = \cos \frac{1}{2}x$
 Give your answers in radians, to two decimal places.

3 Solve the equation
$$3 \sin 2x = \cos x$$
for values of x in the interval $0° \leq \theta \leq 360°$, giving your answers to the nearest $0.1°$, where necessary.

4 a Prove the identity $\tan A + \cot A \equiv 2 \csc 2A$.
 b Hence, for $0 < x < 2\pi$, solve the equation $\tan x + \cot x = 8$, giving your answers in radians to two decimal places.

5 a Express $\cos 2A$ in terms of
 i $\cos A$ **ii** $\sin A$
 b Prove that   **i** $\dfrac{2\cos\theta - \sec\theta}{\csc\theta - 2\sin\theta} \equiv \tan\theta$ **ii** $8\sin^2\left(\dfrac{\theta}{2}\right)\cos^2\left(\dfrac{\theta}{2}\right) \equiv 1 - \cos 2\theta$

6 Solve the equation
$$\tan 2x + \tan x = 0$$
for values of x in the interval $0° \leq x \leq 360°$.

7 A curve is defined by the parametric equations $x = \sin t$, $y = \cos 2t$, $0 \leq t < 2\pi$.
 a Find a Cartesian equation of the curve.
 b Hence find the exact coordinates of the points of intersection of the curve with the coordinate axes.

8 Expand $\cos(2A + A)$ and hence express $\cos 3A$ in terms of $\cos A$.

9 Find

   a $\int \sin^2 4x \, dx$ **b** $\int \cos 3x \sin 3x \, dx$

10 The region enclosed by the curve $y = 2 \sin x$, where $0 \leq x \leq \pi$, and the x-axis is rotated through 2π radians about the x-axis. Find the volume of the solid generated.

11 a Write $\cos^2 A$ in terms of $\cos 2A$.

   b The diagram shows the graph of $y = \cos^2 x$ for $-\frac{1}{2}\pi \leq x \leq \frac{1}{2}\pi$. Find the area enclosed by the curve and the x-axis.

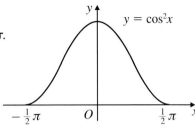

12 Find

   a $\int (\cos x + \sin x)^2 \, dx$ **b** $\int (\cos x + \sin x)(\cos x - \sin x) \, dx$

SKILLS CHECK **4B EXTRA** is on the CD

Examination practice 4: Trigonometry

1 a Given that $\sin \alpha = \frac{12}{13}$, where α is an obtuse angle, find the exact value of $\cos \alpha$.

   b Given also that $\cos \beta = \frac{4}{5}$, where β is an acute angle, find the exact value of $\sin(\alpha + \beta)$.

[AQA Jan 2002]

2 a Show that $\sin(\alpha + \beta) + \sin(\alpha - \beta) \equiv 2 \sin \alpha \cos \beta$.

   b i Express $2 \sin 8x \cos 2x$ in the form $\sin A + \sin B$.

   ii Hence find $\int 6 \sin 8x \cos 2x \, dx$.

[AQA June 2004]

3 a Express $6 \cos x - 8 \sin x$ in the form $R \cos(x + \alpha)$, where R is a positive constant and $0° < \alpha < 90°$. Give the value of α to the nearest $0.1°$.

   b Hence find the solutions, to the nearest degree, of the equation

   $6 \cos x - 8 \sin x = 3$

   where $0° < x < 360°$.

4 a Find the value of $\tan^{-1} 2.4$, giving your answer in radians to three decimal places.

   b Express $10 \sin \theta + 24 \cos \theta$ in the form $R \sin(\theta + \alpha)$, where $R > 0$ and $0 < \alpha < \frac{\pi}{2}$.

   c Hence

   i write down the maximum value of $10 \sin \theta + 24 \cos \theta$;

   ii find a value of θ at which this maximum value occurs.

[AQA Jan 2004]

5 a Express $7 \cos \theta + 24 \sin \theta$ in the form $r \cos(\theta - \alpha)$, where $r > 0$ and $0 < \alpha < \frac{\pi}{2}$.

   b Hence, or otherwise, solve the equation

   $7 \cos \theta + 24 \sin \theta = 12.5$

   for $0 < \theta < 2\pi$, giving your answers to one decimal place.

6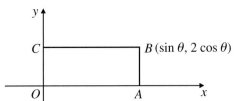

The diagram shows a rectangle $OABC$ in which B has coordinates $(\sin\theta, 2\cos\theta)$ where $0 \leq \theta \leq \frac{\pi}{2}$. The perimeter of the rectangle is of length L.

   a **i** Write down the length L in terms of θ.

      ii Hence obtain an expression for L in the form $R\sin(\theta + \alpha)$, where $R > 0$ and $0 \leq \alpha \leq \frac{\pi}{2}$. Give your answer for α to three decimal places.

   b Given that θ varies between 0 and $\frac{\pi}{2}$:

      i write down the maximum value of L;

      ii find the value of θ, to two decimal places, for which L is maximum.   [AQA Jan 2003]

7 **a** Show that $\dfrac{\cot^2\theta}{1 + \cot^2\theta} \equiv \cos^2\theta$.

   b Hence solve $\dfrac{\cot^2\theta}{1 + \cot^2\theta} = 2\sin 2\theta$ for $0° \leq \theta \leq 360°$.   [AQA June 2002]

8 Solve the equation
$$3\cos 2\theta - \cos\theta + 1 = 0$$
giving all solutions in degrees to the nearest degree in the interval $0° \leq \theta \leq 360°$.

9 Prove the identity
$$\frac{1 - \cos 2\theta}{1 + \cos 2\theta} \equiv \sec^2\theta - 1$$

10 Solve the equation
$$18\tan\theta = \tan 2\theta$$
giving all solutions to the nearest $0.1°$ in the interval $-90° < \theta < 90°$.

11 **a** Express $\sin^2 A$ in terms of $\cos 2A$.

    b Find $\displaystyle\int (3\sin^2 x + 2\cos^2 x)\,dx$.

12 **a** Express $6\sin x\cos x + 4\sin^2 x - 2$ in the form $a\sin 2x + b\cos 2x$, where a and b are constants to be found.

    b Hence solve $6\sin x\cos x + 4\sin^2 x - 2 = 0$ for $0° < x < 360°$.

13 **a** Simplify $\cos 2\theta\cos\theta - \sin 2\theta\sin\theta$.

    b Hence, or otherwise, solve the equation
$$4\cos 2\theta\cos\theta = 4\sin 2\theta\sin\theta + 1,$$
giving all solutions to the nearest degree in the interval $0° < \theta < 180°$.   [AQA (AEB) Jan 1998]

14 a Express $3\cos^2 \tfrac{1}{2}x$ in the form $a + b\cos x$, where a and b are constants.

 b Hence find $\int 3\cos^2 \tfrac{1}{2}x \, dx$.

15 a Express $\sin 2A$ in terms of $\sin A$ and $\cos A$.

 b Express $\cos 2A$ in terms of $\sin A$.

 c Prove the identity
$$\frac{\sin 2A}{1 - \cos 2A} \equiv \cot A$$

16 a Prove the identity
$$(3\sin x + 5\cos x)^2 \equiv 17 + 8\cos 2x + 15\sin 2x$$

 b Hence find:

  i $\int (3\sin x + 5\cos x)^2 \, dx$;

  ii the volume of the solid formed when the region bounded by the curve with equation $y = 3\sin x + 5\cos x$, the coordinate axes and the line $x = \dfrac{\pi}{4}$ is rotated through 2π radians about the x-axis.

 c i Show that the equation
$$(3\sin x + 5\cos x)^2 = 4\cos^2 x$$
can be written in the form
$$(3\tan x + 5)^2 = 4$$

  ii Hence, or otherwise, solve the equation
$$(3\sin x + 5\cos x)^2 = 4\cos^2 x$$
giving all solutions in radians in the interval $0 < x < \pi$. [AQA June 2004]

 17 Use the expansion of $\tan(A + B)$, with $B = 2A$, to show that
$$\tan 3A \equiv \frac{3\tan A - \tan^3 A}{1 - 3\tan^2 A}$$

5 Exponentials and logarithms

5.1 Exponential growth and decay

Exponential growth and decay.

Given positive constants a, b and k

- the function $f(t) = a \times b^{kt}$ can be used to model **exponential growth**
- the function $f(t) = a \times b^{-kt}$ can be used to model **exponential decay**.

The graphs of $y = f(t)$ are as follows:

Exponential growth
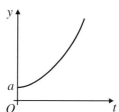

Exponential decay
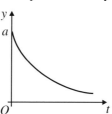

Note:
When $t = 0$, $y = a \times b^0 = a$, since $b^0 = 1$.

Example 5.1 £5000 is invested in a savings account and no further deposits are made. Interest accrues in the account such that the amount in the account after t years is £y, where

$$y = 5000 \times 1.06^t$$

a How much money is in the account after three years?

b After how long has the amount in the account doubled?

Note:
This is an example of exponential growth.

Step 1: Substitute an appropriate value for t.

a $y = 5000 \times 1.06^t$

When $t = 3$,

$$y = 5000 \times 1.06^3 = 5955.08$$

After three years there will be £5955 in the account.

Step 1: Substitute a suitable value for y.

b When $y = 10\,000$,
$$10\,000 = 5000 \times 1.06^t$$

Step 2: Solve for t.
$$1.06^t = \frac{10\,000}{5000}$$
$$= 2$$
$$\log_{10} 1.06^t = \log_{10} 2$$
$$t\log_{10} 1.06 = \log_{10} 2$$
$$t = \frac{\log_{10} 2}{\log_{10} 1.06} = 11.895\ldots$$

The amount in the account will have doubled in approximately 12 years.

Recall:
Exponential equations (C2 Section 4.3).

Recall:
$\log p^q = q \log p$ (C2 Section 4.2).

Note:
You could use logs to the base e (ln).

Example 5.2 The value £V of a motorbike, t years from new, is modelled by the equation

$$V = ap^{-t}$$

where p and a are positive constants.

Note:
This is an example of exponential decay.

A motorbike was purchased for £8000 and after two years its value had decreased to £5120.

a Write down the value of a.

b Find the value of p.

Step 1: Substitute an appropriate value of t.

a When $t = 0$,
$$8000 = ap^0$$
$$a = 8000$$

Tip: $p^0 = 1$, so a is the initial cost.

Step 1: Substitute the given values into the equation.

b When $t = 2$, $V = 5120$ and $a = 8000$
$$\Rightarrow 5120 = 8000 \times p^{-2}$$

Step 2: Rearrange to find p.
$$\frac{5120}{8000} = \frac{1}{p^2}$$
$$p^2 = \frac{8000}{5120} = 1.5625$$
$$p = \sqrt{1.5625} = 1.25$$

Tip: Since p is positive, you need the positive value of the square root.

Rate of growth and rate of decay

If $y = f(t)$, then the **rate of growth or decay** is given by $\dfrac{dy}{dt}$.

Recall: The exponential function e^t (C3 Section 3.1).

You will need to be able to find this in the following *special case* of exponential growth or decay, modelled by the formulae
$$y = Ae^{kt} \quad \text{or} \quad y = Ae^{-kt}$$
where A and k are positive constants.

Recall: $\dfrac{d}{dt}(e^{kt}) = ke^{kt}$ (C3 Section 4.2).

So,
- when $y = Ae^{kt}$, the rate of growth is given by $\dfrac{dy}{dt} = Ake^{kt}$
- when $y = Ae^{-kt}$, the rate of growth is given by $\dfrac{dy}{dt} = -Ake^{-kt}$

Note: The negative value for $\dfrac{dy}{dt}$ represents decay.

Example 5.3 The temperature, $T\,°C$, of a microwave meal t minutes after it has been heated is given by
$$T = 18 + 50e^{-\frac{t}{20}}, \quad t \geqslant 0$$

a Find, in °C, the temperature of the meal the instant that it has been heated.

b Calculate, in °C to three significant figures, the temperature of the meal 5 minutes after it has been heated.

c Calculate, to the nearest minute, the time at which the temperature of the meal is 45 °C.

d Find the rate at which the meal is losing heat 10 minutes after it is taken from the oven.

Step 1: Substitute $t = 0$ to find T.

a When $t = 0$,
$$T = 18 + 50e^{-\frac{0}{20}}$$
$$= 18 + 50e^0$$
$$= 18 + 50 \times 1$$
$$= 68$$

Tip: t is the time after the meal has been heated, so $t = 0$ when the meal comes out of the oven.

Recall: $e^0 = 1$ (C3 Section 3.1).

So the temperature of the meal the instant that it has been heated is 68 °C.

Step 1: Substitute the given value of *t* to find *T*.	**b** When $t = 5$, $$T = 18 + 50\,e^{-\frac{5}{20}}$$ $$= 18 + 50\,e^{-\frac{1}{4}}$$ $$= 18 + 50 \times 0.7788\ldots$$ $$= 56.94\ldots$$ So the temperature of the meal five minutes after it has been heated is 56.9 °C (3 s.f.).	**Tip:** Use the full value in your calculation, but don't forget to give your answer to the required degree of accuracy.
Step 1: Substitute the given value of *T* into the equation.	**c** When $T = 45$, $$45 = 18 + 50\,e^{-\frac{t}{20}}$$ $$27 = 50\,e^{-\frac{t}{20}}$$	
Step 2: Simplify the equation.	$$\frac{27}{50} = e^{-\frac{t}{20}}$$	**Note:** You need to isolate the exponential term ($e^{-\frac{t}{20}}$) before you take logs.
Step 3: Take natural logs of both sides and use an appropriate log law to simplify.	$$\ln \frac{27}{50} = \ln e^{-\frac{t}{20}}$$ $$= -\frac{t}{20}\ln e$$	**Recall:** $\ln e = 1$ (C3 Section 3.2).
Step 4: Rearrange to find *t* and evaluate using your calculator.	$$= -\frac{t}{20}$$ $$-20\ln \frac{27}{50} = t$$ $$t = 12.3\ldots$$ So the temperature of the meal is 45 °C approximately 12 minutes after it is heated.	**Tip:** Use the exact values in your working; only use decimals at the end.
Step 1: Differentiate *x* with respect to *t*.	**d** $T = 18 + 50\,e^{-\frac{t}{20}}$ $$\frac{dT}{dt} = -\tfrac{1}{20} \times 50\,e^{-\frac{t}{20}}$$	**Recall:** $\frac{d}{dx}(e^{kx}) = ke^{kx}$ (C3 Section 4.2)
Step 2: Substitute the given value of *t*.	When $t = 10$, $$\frac{dT}{dt} = -\tfrac{1}{20} \times 50\,e^{-\frac{10}{20}} = -1.516\ldots$$ So, ten minutes after it is taken from the oven, the meal is losing heat at approximately 1.5 °C per minute.	**Note:** Since $\frac{dT}{dt} < 0$, the meal is losing heat.

Example 5.4 In a particular situation, exponential growth is modelled by the formula $x = Ae^{kt}$, where A and k are positive constants. Show that the rate of growth of x is proportional to x.

Step 1: Differentiate *x* with respect to *t*.	$x = Ae^{kt} \Rightarrow \dfrac{dx}{dt} = Ake^{kt}$	**Recall:** The rate of growth is given by $\frac{dx}{dt}$.
Step 2: Write in terms of *x*.	Since $Ae^{kt} = x$, $$\frac{dx}{dt} = k(Ake^{kt})$$	**Note:** This is a very important result and is often used in reverse (Section 6.1).
	$$= kx$$ i.e. $\dfrac{dx}{dt} \propto x$ Hence the rate of growth of *x* is proportional to *x*.	

SKILLS CHECK 5A: Exponential growth and decay

 1 In a particular model for exponential growth, $y = a \times 10^{kt}$, where a and k are positive constants. It is known that $y = 2$ when $t = 0$ and $y = 4$ when $t = 1$.

   a Find the value of a.

   b Find the exact value of k.

   c Find the value of y when $t = 3$.

2 It is known that, in a particular model for exponential decay,
$$x = 3e^{-5t}$$

   a Find $\dfrac{dx}{dt}$ **i** in terms of t **ii** in terms of x.

   b Find $\dfrac{dx}{dt}$ when $t = 4$.

   c Find $\dfrac{dx}{dt}$ when $x = 4$.

3 The number of bacteria present in a culture at a time t hours after the beginning of an experiment is given by $y = Ae^{2t}$.

   a Find the number of bacteria after one hour if there are 1000 bacteria at the beginning of the experiment.

   b Sketch the graph of y against t showing the coordinates of any intercepts with the axes.

   c i Find the rate at which the number of bacteria is increasing after one hour.

       ii Explain how this can be illustrated on the sketch.

4 A model giving the number, N, of micro-organisms present at time t hours after the start of an experiment is given by
$$N = 200e^{0.018t}$$

   a Find the value of t for which the number of micro-organisms is 2000.

   b Find the rate at which the number of micro-organisms is increasing when $t = 20$.

5 The amount an initial investment of £1000 is worth after t years is given by A, where
$$A = 1000\,e^{0.09t}, \quad t \geq 0$$

   a How much is the investment worth after five years?

   b After how many years will the investment have doubled in value?

 6 The mass, in grams, of a radioactive substance after t years, where $t \geq 0$, is given by
$$m = 50e^{-kt}$$
where k is a positive constant.

   a What is the value of m when $t = 0$?

   After 300 years, the mass will be half its value when $t = 0$.

   b Find the value of k, to two significant figures.

   c Find the rate at which the mass is decreasing when $t = 250$.

SKILLS CHECK 5A EXTRA is on the CD

Examination practice 5: Exponentials and logarithms

1. The amount of money, £P, in a special savings account at time t years after 1st January 2000 is given by

   $P = 100 \times 1.05^t$

   a State the amount of money in the account on 1st January 2000.

   b Calculate, to the nearest penny, the amount of money in the account on 1st January 2004.

   c Find the value of t when $P = 150$, giving your answer to 3 significant figures. [AQA June 2004]

2. a On 1 January 1998, Company A issued some shares at a price of 90p each. Investors expected the shares to increase in value according to the model

   $P = 90 \times 1.12^t$,

   where the value of each share after t years is P pence.

   Show that, according to this model, the value of each share on 1 January 2005, correct to the nearest penny, is 199p.

   b On 1 January 1998, Company B issued some shares at a price of 270p each. On 1 January 2005, these shares are worth 405p each. The value, Q pence, of each share t years after their issue can be represented by the model

   $Q = 270 \times k^t$,

   where k is a constant.

   Show that the value of k, correct to two decimal places, is 1.06.

   c Assuming that the two models remain valid, there is a time at which a share in Company A will be equal in value to a share in Company B.

   Determine the year in which this happens. [AQA Jan 2005]

3. A microbiologist is studying the growth of populations of simple organisms.

   For one such organism, the model proposed is

   $P = 100 - 50\mathrm{e}^{-\frac{1}{4}t}$,

   where P is the population after t minutes.

   a Write down:

   i the initial value of the population;

   ii the value which the population approaches as t becomes large.

   b Find the time at which the population will have a value of 75, giving your answer to two significant figures. [AQA Jan 2004]

4. A biologist is studying the growth of a population of rabbits. A proposed model for the size of the population, P rabbits, t months after the study started is

   $P = 20\mathrm{e}^{\left(\frac{t-6}{4}\right)}$.

   a Use this model to find, to the nearest whole number, the size of the population:

   i after 6 months;

   ii after 12 months.

   b Find the time, in months, when the population first exceeds 1000 rabbits. [AQA June 2005]

5 A model for the temperature, $T°$ celsius, of a cup of tea x minutes after being poured from a teapot is given by

$$T = 63 \times 3^{-\left(\frac{x}{A}\right)} + 21,$$

where A is a positive constant.

a Use the model to:
  i find the temperature of the cup of tea immediately after being poured;
  ii predict the temperature after the cup of tea has been standing a very long time.

b By rearranging the equation and taking natural logarithms, show that

$$A \ln\left(\frac{63}{T - 21}\right) = x \ln 3.$$

c Given that $T = 40$ when $x = 15$, find the value of A, giving your answer to three significant figures. [AQA June 2002]

6 The height of a boy was recorded for the first seven years of his life. A model suggested for the boy's height, y cm, when he was x years old is given by

$$y = 130 - 80e^{-\frac{5x}{16}}, \quad 0 \leqslant x \leqslant 7.$$

a Use this model to estimate:
  i the height of the boy at birth;
  ii the age of the boy, to the nearest month, when his height was 1 metre;
  iii the rate of growth, in cm per year, when the boy was exactly 4 years old.

b Explain why this model might be unsuitable for predicting the boy's height when he is a teenager. [AQA Jan 2002]

7 a Given that $N = Ae^{kt}$, where A and k are constants, show that $\dfrac{dN}{dt} = kN$.

b The number of bacteria, N, in a colony is such that the rate of increase of N is proportional to N. The time, t, is measured in hours from the instant that $N = 2$ million. When $t = 3$, $N = 5$ million. Find the value of t when $N = 8$ million. [AQA Jan 2002]

6 Differentiation and integration

6.1 Forming differential equations

Formation of simple differential equations.

The differential coefficient $\frac{dy}{dx}$ is known as the first order derivative of y with respect to x.

An equation containing $\frac{dy}{dx}$, but not any higher order derivatives such as $\frac{d^2y}{dx^2}$, is called a **first order differential equation**.

Examples of first order differential equations are

$$\frac{dy}{dx} = 2x, \quad y\frac{dy}{dx} = e^x + 2, \quad \frac{dy}{dx} = -y$$

Differential equations are often used to model economic, social or scientific situations and many scientific laws can be expressed as differential equations.

Note: $\frac{dy}{dx}$ gives the rate of change of y with respect to x.

Note: The rate of change is often required with respect to time t, for example $\frac{dV}{dt} = \frac{1}{2}V$.

Example 6.1 A mathematical model for the number of bacteria, N, in an experiment, states that N is increasing at a rate proportional to the number of bacteria present at time t. Write down a differential equation involving N and t.

Step 1: Use the given information to form a differential equation.

The rate of change of N is $\frac{dN}{dt}$.

So $\frac{dN}{dt} \propto N$

i.e. $\frac{dN}{dt} = kN$, where k is a positive constant

Note: To find the value of the proportionality constant k you would need more information.

Example 6.2 According to Newton's law of cooling, the rate of temperature loss of a body is proportional to the difference between the temperature of the body and the temperature of the surrounding air. Write a differential equation for this law.

Step 1: Define the variables.

Let T be the temperature of the body at time t seconds and let T_0 be the temperature of the surrounding air.

Step 2: Use the given information to form a differential equation.

The rate of change of T is given by $\frac{dT}{dt}$,

so $\frac{dT}{dt} \propto (T - T_0)$

Since the body is losing heat,

$\frac{dT}{dt} = -k(T - T_0)$, where k is a positive constant

Tip: $T - T_0$ is the difference between the temperature of the body and the temperature of the surrounding air.

Tip: Use the negative sign, with positive proportionality constant k, to show that the temperature is decreasing.

Example 6.3 At time t, the surface area of a circular oil slick with radius r is increasing at a constant rate of 50 m²/s.

Find a differential equation for the rate of change of the radius in terms of r and t.

Step 1: Define the variables.

Let the surface area of the oil slick at time t seconds be A m^2.

Step 2: Express the given rate of change in calculus form.

A is increasing at a rate of 50 m^2/s $\Rightarrow \dfrac{dA}{dt} = 50$

Step 3: Express A in terms of r and differentiate with respect to r.

$A = \pi r^2 \Rightarrow \dfrac{dA}{dr} = 2\pi r$

We require the rate of change of r, i.e. $\dfrac{dr}{dt}$.

Step 4: Use the chain rule to differentiate r with respect to t.

By the chain rule,
$$\dfrac{dr}{dt} = \dfrac{dr}{dA} \times \dfrac{dA}{dt}$$

Since $\dfrac{dA}{dr} = 2\pi r$, $\dfrac{dr}{dA} = \dfrac{1}{2\pi r}$

Step 5: Substitute known values.

So $\dfrac{dr}{dt} = \dfrac{1}{2\pi r} \times 50$

$\dfrac{dr}{dt} = \dfrac{25}{\pi r}$

Recall: Chain rule (C3 Section 4.2).

Recall: $\dfrac{dr}{dA} = \dfrac{1}{\frac{dA}{dr}}$ (C3 Section 4.3).

6.2 Solving differential equations

Analytical solution of simple first order differential equations with separable variables.

You solved the following type of first order differential equation in previous modules:

$$\text{If } \dfrac{dy}{dx} = f(x), \text{ then } y = \int f(x)\, dx$$

For example:

$$\text{If } \dfrac{dy}{dx} = 2x + 1, \text{ then } y = \int (2x + 1)\, dx = x^2 + x + c$$

This solution contains an integration constant (c) and is called the **general solution** of the differential equation.

If, however, we know that $y = 10$ when $x = 2$, we can find a **particular solution** as follows:

$y = 10$ when $x = 2 \Rightarrow 10 = 2^2 + 2 + c$

$c = 4$

Hence the particular solution is $y = x^2 + x + 4$.

In *Core 4* the differential equations that you have to solve include ones such as

$$\dfrac{dy}{dx} = 3x^2 y \quad \text{or} \quad e^x \dfrac{dy}{dx} = \dfrac{1}{y}$$

The expressions in x and y will be able to be separated into the form

$$f(y) \dfrac{dy}{dx} = g(x)$$

Recall: Integration (C1 Chapter 4, C2 Chapter 6 and C3 Chapter 5).

Note: A particular solution does not have an integration constant.

Note: This is known as 'variables separable' form.

To solve, integrate both sides with respect to x to give

$$\int f(y) \frac{dy}{dx} dx = \int g(x) dx$$

This reduces to the form

$$\int f(y) dy = \int g(x) dx$$

Note:
In practice, take all the terms in y to the side with dy and take all the terms in x to the other side with dx. This is known as **separating the variables**.

Example 6.4 Given that $\frac{dy}{dx} = \frac{x}{y^2}$, express y in terms of x.

Step 1: Separate the variables to get the form $\int f(y) dy = \int g(x) dx$.

$$\frac{dy}{dx} = \frac{x}{y^2}$$

$$\Rightarrow \int y^2 dy = \int x\, dx$$

Step 2: Integrate each side separately.

$$\Rightarrow \tfrac{1}{3} y^3 = \tfrac{1}{2} x^2 + c$$

Step 3: Make y the subject.

$$y^3 = \tfrac{3}{2} x^2 + d$$

$$y = \sqrt[3]{\tfrac{3}{2} x^2 + d}$$

Note:
$y^2 \frac{dy}{dx} = x$
so $\int y^2 \frac{dy}{dx} dx = \int x\, dx$.

Tip:
You can use any letter for the constant. Here $d = 3c$.

Example 6.5 It is given that $\frac{dy}{dx} = 3x^2 y$.

a Show that $y = Ae^{x^3}$, where A is a constant.

b Given also that $y = 2$ when $x = 0$, find the value of A.

Step 1: Separate the variables to get the form $\int f(y) dy = \int g(x) dx$.

a
$$\frac{dy}{dx} = 3x^2 y$$

$$\Rightarrow \int \frac{1}{y} dy = \int 3x^2 dx$$

Step 2: Integrate each side separately.

$$\Rightarrow \ln y = x^3 + c$$

Step 3: Make y the subject and write in the required form.

$$y = e^{x^3 + c}$$
$$= e^{x^3} \times e^c$$
$$= Ae^{x^3}$$

Alternatively:
$$\ln y = x^3 + c$$
$$\ln y = x^3 + \ln A$$
$$\ln y - \ln A = x^3$$
$$\ln \frac{y}{A} = x^3$$
$$y = Ae^{x^3}$$

Step 4: Use the given condition to find the value of the constant.

b $y = 2$ when $x = 0$
$$\Rightarrow 2 = Ae^0$$
$$2 = A \times 1$$

Hence $A = 2$.

Note:
It is better to leave numerical factors where they will be in the numerator of a fraction.

Recall:
$\ln a = b \Leftrightarrow e^b = a$
(C3 Section 3.2).

Tip:
Let $e^c = A$.

Tip:
Let $c = \ln A$.

Recall:
$\log a - \log b = \log \frac{a}{b}$
(C2 Section 4.2).

Recall:
$e^0 = 1$

The differential equation may be set in the context of a problem, as in the following example.

Example 6.6 A mathematical model for the number of bacteria, N, in an experiment states that N is increasing at a rate proportional to the number of bacteria present at time t, where t is measured in minutes. Initially there are 1000 bacteria and after 5 minutes there are 10 000.

a Show that $N = 1000e^{kt}$ and find the exact value of k in the form $a \ln b$.

b Calculate the number of bacteria after 8 minutes, giving your answer to two significant figures.

Note: This is the situation described in Example 6.1.

Step 1: Use the given information to form a differential equation.

a The rate of change of N is $\dfrac{dN}{dt}$.

$$\dfrac{dN}{dt} \propto N \Rightarrow \dfrac{dN}{dt} = kN, \text{ where } k \text{ is a positive constant}$$

Note: k is the proportionality constant.

Step 2: Separate the variables and integrate.

$$\dfrac{dN}{dt} = kN$$

$$\Rightarrow \int \dfrac{1}{N} dN = \int k \, dt$$

$$\Rightarrow \ln N = kt + c$$

Step 3: Substitute the first given condition to find c.

When $t = 0$, $N = 1000$.

So $\quad \ln 1000 = k \times 0 + c$

$\quad\quad c = \ln 1000$

Hence $\quad \ln N = kt + \ln 1000$

$\ln N - \ln 1000 = kt$

$\ln \dfrac{N}{1000} = kt$

$\dfrac{N}{1000} = e^{kt}$

$N = 1000e^{kt}$

Note: Knowing the initial number of bacteria enables the integration constant to be found.

Tip: As in Example 6.5 you could write $N = Ae^{kt}$ straight away and then find A.

Step 4: Substitute the second given condition to find k.

When $t = 5$, $N = 10\,000$.

So $\quad 10\,000 = 1000e^{k \times 5}$

$e^{5k} = 10$

$5k = \ln 10$

$k = \tfrac{1}{5} \ln 10$

Note: Knowing a second condition enables the proportionality constant to be found.

Recall: $e^b = a \Leftrightarrow \ln a = b$

Step 5: Write out the particular solution and substitute $t = 8$.

b $N = 1000e^{(\frac{1}{5}\ln 10) \times t}$

When $t = 8$,

$N = 1000e^{(\frac{1}{5}\ln 10) \times 8}$

$= 1000e^{\frac{8}{5}\ln 10}$

$= 39\,810.7\ldots$

$= 40\,000$ (2 s.f.)

After 8 minutes, there are approximately 40 000 bacteria.

SKILLS CHECK 6A: First order differential equations

1. Express y in terms of x when

   a $\dfrac{dy}{dx} = y \cos x$

   b $\dfrac{dy}{dx} = 4x\sqrt{y}$

   c $\dfrac{dy}{dx} = \dfrac{y}{x}$

2. Given that $2\dfrac{dx}{dt} = \dfrac{e^t}{x}$ and that $x = 2$ when $t = 0$, express t in terms of x.

3. **a** Obtain the general solution of the differential equation $\dfrac{dy}{dx} = x(y + 1)$.

   b Given that $y = 2$ when $x = 0$, find the particular solution, expressing y as a function of x.

4. **a** Obtain the general solution of the differential equation $\dfrac{dy}{dx} = \tfrac{1}{2}y^3 x^2$.

   b Given also that $y = 1$ at $x = 3$, show that $y^2 = \dfrac{3}{30 - x^3}$, $x \neq \sqrt[3]{30}$.

5. A curve has equation $y = f(x)$. The gradient of the curve at the point (x, y) is given by

   $$\dfrac{dy}{dx} = \dfrac{2x}{1 - 2y}$$

   Given also that $(0, 2.5)$ lies on the curve, show that the curve is a circle and find the coordinates of the centre of the circle, and its radius.

6. Given that $x = 1$ when $t = 1$, express x in terms of t, where

   a $\dfrac{dx}{dt} = \dfrac{x + 3}{4t}$

   b $\dfrac{dx}{dt} = \dfrac{x + 3}{4}$

7. Find the general solution of the differential equation $\dfrac{d\theta}{dx} = x \cos^2 \theta$.

8. **a** Use integration by parts to find $\int x \cos 2x \, dx$.

   b **i** Express $\cos^2 x$ in terms of $\cos 2x$.

   ii Find $\int \cos^2 x \, dx$.

   c Given that $y = 0$ at $x = \tfrac{1}{4}\pi$, solve the differential equation $\dfrac{dy}{dx} = \dfrac{x \cos 2x}{\cos^2 y}$.

9. **a** Using the substitution $u = 9 + x^3$, find $\int x^2 (9 + x^3)^4 \, dx$.

   b Given that $y = 0$ when $x = -2$, solve the differential equation $\dfrac{dy}{dx} = \dfrac{15x^2(9 + x^3)^4}{e^y}$, expressing your answer in the form $y = \ln f(x)$.

10 According to Newton's law of cooling, the rate of temperature loss of a body is proportional to the difference between the temperature of the body, T, and the temperature of the surrounding air. The air in a room has a constant temperature of 15 °C.

   a Show that $\dfrac{dT}{dt} = -k(T - 15)$, where k is a positive constant.

   b Show, by integration, that $T = 15 + Ae^{-kt}$, where A is a constant.

   The temperature of an object in the room is found to be 75 °C. Ten minutes later its temperature has dropped by 10 °C.

   c Show that $k = \frac{1}{10} \ln \left(\frac{6}{5}\right)$.

   d Find the temperature of the object after a further ten minutes.

SKILLS CHECK **6A EXTRA** is on the CD

6.3 Differentiation of implicit and parametric functions

Differentiation of simple functions defined implicitly or parametrically. Equations of tangents and normals for curves specified implicitly or in parametric form.

Implicit functions

In *Core 3* the functions that you differentiated were of the form $y = f(x)$, where y was given explicitly in terms of x, or $x = f(y)$, where x was given explicitly in terms of y.

Sometimes a function in two variables is defined **implicitly**, and neither variable is given explicitly in terms of the other.

You will need to be able to differentiate implicit functions such as $3x^2 + y^2 = 9$ and $x^2y + y^3 = 2x$.

To differentiate an implicit function:

- differentiate term by term with respect to x
- use the chain rule to differentiate any terms in y.

Recall:
The chain rule (C3 Section 4.2).

For example, to differentiate y^2 with respect to x use the chain rule:

$$\frac{d}{dx}(y^2) = \frac{d}{dy}(y^2) \times \frac{dy}{dx}$$
$$= 2y\frac{dy}{dx}$$

Tip:
When you have understood implicit differentiation you will be able to show far less working.

Example 6.7 Find $\dfrac{dy}{dx}$ in terms of x and y for each of the following:

   a $3x^2 + y^2 = 9$

   b $x^2y + y^3 = 2x$

Step 1: Differentiate term by term with respect to x.

a $3x^2 + y^2 = 9$

Differentiating with respect to x,

$$\frac{d}{dx}(3x^2) + \frac{d}{dx}(y^2) = \frac{d}{dx}(9)$$

Tip: Don't forget any terms on the right-hand side.

Step 2: Use the chain rule to differentiate y^2 with respect to x.

$$6x + \frac{d}{dy}(y^2)\frac{dy}{dx} = 0$$

$$6x + 2y\frac{dy}{dx} = 0$$

Step 3: Rearrange to make $\frac{dy}{dx}$ the subject.

$$2y\frac{dy}{dx} = -6x$$

$$\frac{dy}{dx} = -\frac{6x}{2y}$$

$$= -\frac{3x}{y}$$

Tip: Divide through by the coefficient of $\frac{dy}{dx}$.

Step 1: Differentiate term by term with respect to x.

b $x^2y + y^3 = 2x$

Differentiating with respect to x,

$$\frac{d}{dx}(x^2y) + \frac{d}{dx}(y^3) = \frac{d}{dx}(2x)$$

Step 2: Use the product rule to differentiate x^2y with respect to x.

$$\left(\frac{d}{dx}(x^2) \times y + x^2 \times \frac{d}{dx}(y)\right) + \frac{d}{dx}(y^3) = 2$$

Recall: Product rule (C3 Section 4.3).

Step 3: Use the chain rule to differentiate y and y^3 with respect to x.

$$2xy + x^2\frac{d}{dy}(y)\frac{dy}{dx} + \frac{d}{dy}(y^3)\frac{dy}{dx} = 2$$

$$2xy + x^2\frac{dy}{dx} + 3y^2\frac{dy}{dx} = 2$$

Tip: $\frac{d}{dy}(y) = 1$

$$(x^2 + 3y^2)\frac{dy}{dx} = 2 - 2xy$$

Tip: Collect all terms containing $\frac{dy}{dx}$ on one side and factorise.

Step 4: Rearrange to make $\frac{dy}{dx}$ the subject.

$$\frac{dy}{dx} = \frac{2 - 2xy}{x^2 + 3y^2}$$

Example 6.8 A circle has centre (2, 3) and radius $\sqrt{5}$.

  a State the equation of the circle.

  b Use implicit differentiation to find an equation of the normal to the circle at (1, 5).

Note: Although you learnt another method for answering **b** in C1, this question tells you to use implicit differentiation.

Step 1: Use the general equation of a circle.

a The circle with centre (2, 3) and radius $\sqrt{5}$ has equation $(x - 2)^2 + (y - 3)^2 = 5$.

Recall: A circle, centre (a, b), radius r, has equation $(x - a)^2 + (y - b)^2 = r^2$ (C1 Section 2.3).

Step 2: Differentiate term by term with respect to x.

b $(x - 2)^2 + (y - 3)^2 = 5$

Differentiating with respect to x,

$$\frac{d}{dx}(x-2)^2 + \frac{d}{dx}(y-3)^2 = \frac{d}{dx}(5)$$

Step 3: Use the chain rule to differentiate $(y-3)^2$ with respect to x.

$$2(x-2) + \frac{d}{dy}(y-3)^2 \frac{dy}{dx} = 0$$

$$2x - 4 + 2(y-3)\frac{dy}{dx} = 0$$

Step 4: Substitute the given point and rearrange to find the gradient of the tangent.

At $(1, 5)$,

$$2 - 4 + 2(5-3)\frac{dy}{dx} = 0$$

$$4\frac{dy}{dx} = 2$$

$$\frac{dy}{dx} = \tfrac{1}{2}$$

Step 5: Find the gradient of the normal at the given point.

The gradient of the normal at $(1, 5)$ is -2.

An equation of the normal to the circle at $(1, 5)$ is

Step 6: Use an appropriate straight line equation.

$$y - 5 = -2(x - 1)$$
$$y = 7 - 2x$$

Tip:
By the chain rule
$\frac{d}{dx}(ax+b)^n = an(ax+b)^{n-1}$.

Tip:
It's easier to substitute and then rearrange. As $\frac{dy}{dx}$ is in terms of x and y, you'll need to substitute both the x- and y-coordinates of the point.

Recall:
Tangents and normals are perpendicular, so the product of their gradients is -1 (C1 Section 2.2).

Recall:
Equation of a straight line (C1 Section 2.1).

Parametric functions

To differentiate a function where x and y are defined in terms of the parameter t:

- differentiate x with respect to t to get $\frac{dx}{dt}$
- differentiate y with respect to t to get $\frac{dy}{dt}$
- then use the chain rule to find $\frac{dy}{dx}$.

You can use the format of the chain rule learnt in *Core 3*:

$$\frac{dy}{dx} = \frac{dy}{dt} \times \frac{dt}{dx}$$

Alternatively, since $\frac{dt}{dx} = \frac{1}{\frac{dx}{dt}}$,

$$\frac{dy}{dx} = \frac{dy}{dt} \times \frac{1}{\frac{dx}{dt}}$$

i.e. $\frac{dy}{dx} = \frac{dy}{dt} \div \frac{dx}{dt}$

This alternative format is often a good one to use when a curve is defined by parametric equations.

Recall:
Chain rule (C3 Section 4.2).

Note:
Multiplying by $\frac{dt}{dx}$ is the same as dividing by $\frac{dx}{dt}$.

Example 6.9 A curve has parametric equations $x = 2t^2$, $y = t^3$.

    a Find $\dfrac{dy}{dx}$, leaving your answer in terms of the parameter t.

    b Find the gradient of the curve at the point $(32, -64)$.

Step 1: Differentiate x and y with respect to t.

a $x = 2t^2$ $y = t^3$

$\dfrac{dx}{dt} = 4t$ $\dfrac{dy}{dt} = 3t^2$

Step 2: Use the chain rule to find $\dfrac{dy}{dx}$.

$\dfrac{dy}{dx} = \dfrac{dy}{dt} \div \dfrac{dx}{dt} = \dfrac{3t^2}{4t} = \dfrac{3t}{4}$

Tip:
You could have found the reciprocal of $\dfrac{dx}{dt}$ and used $\dfrac{dy}{dx} = \dfrac{dy}{dt} \times \dfrac{dt}{dx}$, but the new format of the chain rule is easier to apply here.

Step 3: Find the value of t at the given point.

b When $y = -64$,

$y = t^3 \Rightarrow -64 = t^3$

$t = -4$

Tip:
It is better to substitute for y here. Using the x-coordinate will give you two possible values of t, only one of which is the required value.

Step 4: Find $\dfrac{dy}{dx}$ at the given point.

When $t = -4$,

$\dfrac{dy}{dx} = \dfrac{3t}{4} = \dfrac{3 \times (-4)}{4} = -3$

The gradient of the curve at the point $(32, -64)$ is -3.

Example 6.10 A curve is defined by the parametric equations

$$x = 1 - t^3, \quad y = 3t^2$$

Find the equation of the normal to the curve at the point where $t = -1$, in the form $ax + by + c = 0$, where a, b and c are integers.

Step 1: Differentiate x and y with respect to t.

$x = 1 - t^3$ $y = 3t^2$

$\dfrac{dx}{dt} = -3t^2$ $\dfrac{dy}{dt} = 6t$

Note:
You could find the Cartesian equation and differentiate that, but the working is more complicated in this case.

Step 2: Use the chain rule to find $\dfrac{dy}{dx}$.

$\dfrac{dy}{dx} = \dfrac{dy}{dt} \div \dfrac{dx}{dt} = -\dfrac{6t}{3t^2} = -\dfrac{2}{t}$

Step 3: Find $\dfrac{dy}{dx}$ at the given point.

When $t = -1$,

$\dfrac{dy}{dx} = -\dfrac{2}{t} = -\dfrac{2}{-1} = 2$

Step 4: Find the gradient of the normal at the point.

At $t = -1$,

gradient of tangent $= 2 \Rightarrow$ gradient of normal $= -\dfrac{1}{2}$

Recall:
For perpendicular lines, the product of the gradients is -1. (C1 Section 2.2).

Step 5: Substitute the value of t to find the Cartesian coordinates of the point.

When $t = -1$,

$x = 1 - t^3 = 1 - (-1)^3 = 2$

$y = 3t^2 = 3 \times (-1)^2 = 3$

Note:
The Cartesian coordinates are the (x, y) coordinates.

Step 6: Use an appropriate straight line equation.

Equation of normal at $(2, 3)$:

$y - 3 = -\tfrac{1}{2}(x - 2)$

$2(y - 3) = -x + 2$

$2y - 6 = -x + 2$

Step 7: Write in the required format.

$x + 2y - 8 = 0$

Example 6.11 A curve is defined by the parametric equations $x = 3 \sin \theta + 2$, $y = \cos 2\theta + 4$, where $-\frac{1}{2}\pi \leq \theta \leq \frac{1}{2}\pi$.

Note: The parameter is θ.

Find the coordinates of the stationary point on the curve.

Step 1: Differentiate x and y with respect to θ.

$x = 3 \sin \theta + 2 \qquad y = \cos 2\theta + 4$

$\dfrac{dx}{dt} = 3 \cos \theta \qquad \dfrac{dy}{dt} = -2 \sin 2\theta$

Recall: $\dfrac{d}{d\theta}(\cos k\theta) = -k \sin k\theta$ (C3 Section 4.2).

Step 2: Use the chain rule to find $\dfrac{dy}{dx}$.

$\dfrac{dy}{dx} = \dfrac{dy}{dt} \div \dfrac{dx}{dt} = -\dfrac{2 \sin 2\theta}{3 \cos \theta}$

$= -\dfrac{2 \times 2 \sin \theta \cos \theta}{3 \cos \theta}$

$= -\dfrac{4 \sin \theta}{3}$

Recall: $\sin 2\theta \equiv 2 \sin \theta \cos \theta$ (Section 4.3).

Step 3: Set $\dfrac{dy}{dx} = 0$ and solve for θ.

At the stationary point,

$\dfrac{dy}{dx} = 0 \Rightarrow -\dfrac{4 \sin \theta}{3} = 0$

$\sin \theta = 0$

$\theta = 0$

Recall: Stationary points have zero gradient (C2 Section 5.3).

Tip: Here there's only one value of θ in the range, but don't forget to solve for all possible values.

Step 4: Substitute the value of θ to find the x- and y-coordinates.

When $\theta = 0$,

$x = 3 \sin \theta + 2 = 3 \sin 0 + 2 = 2$

$y = \cos 2\theta + 4 = \cos 0 + 4 = 5$

There is a stationary point at $(2, 5)$.

SKILLS CHECK 6B: Differentiation of implicit and parametric functions

1 Find an expression for $\dfrac{dy}{dx}$ in terms of x and y for each of the following curves:

  a $3x^2 - y^2 + 5x - 6y + 5 = 0$

  b $y^3 + x^2y - 2x = 0$

  c $y^3 + x \ln y = 3x^2$

2 Find the coordinates of the stationary points on the curve $x^2 + y^2 - 6x - 8y + 24 = 0$.

3 A curve has equation $x^2 + 3y^2 - 4x - 6y - 14 = 0$.

  a Use implicit differentiation to find $\dfrac{dy}{dx}$.

  b Find an equation of the normal to the curve at the point $(5, -1)$.

4 The curves $xy = 2$ and $y^2 - x^2 = 3$ intersect at the point $(1, 2)$.

  a Find the gradients of the tangents to the curves at this point of intersection.

  b Determine the angle between these tangents.

5 Find an expression for $\dfrac{dy}{dx}$ in terms of t for each of the following curves:

  a $x = 2t^3,\ y = 3t^2 + 1$ **b** $x = 10 \sec t,\ y = 5 \tan t$ **c** $x = t^2 - 4t,\ y = t \ln 4t$

6 A curve has parametric equations $x = t^2 + 1$, $y = 4(t + 1)$.

  a Find $\dfrac{dy}{dx}$ in terms of t.

  The normal to the curve at $(2, 8)$ cuts the x-axis at $(p, 0)$ and the y-axis at $(0, q)$.

  b Find an equation of the normal at $(2, 8)$.

  c Find p and q.

7 A curve has parametric equations $x = 6t + 3 \sin 2t$, $y = 2 \cos 2t$, $0 < t < \dfrac{\pi}{2}$.
Show that $\dfrac{dy}{dx} = -\tfrac{2}{3} \tan t$.

8 A curve has parametric equations $x = at$, $y = at^2 + at$, where a is a constant.
Given that the curve has a stationary point at $(-2, -1)$, find a.

9 A curve has parametric equations $x = e^{2t} + 2t$, $y = e^t - t$.
Find the coordinates of the stationary point on the curve.

SKILLS CHECK 6B EXTRA is on the CD

6.4 Integration using partial fractions

Simple cases of integration using partial fractions.

When you are asked to integrate a rational function where the denominator factorises, a useful method to investigate is whether it can be split into partial fractions. If so, the function may be easier to integrate in this partial fraction form.

Recall:
Partial fractions (Section 1.3).

In Example 1.10 it was shown that
$$\dfrac{2x - 4}{(x - 3)(x + 1)} \equiv \dfrac{1}{2(x - 3)} + \dfrac{3}{2(x + 1)}.$$

So, to find $\displaystyle\int \dfrac{2x - 4}{(x - 3)(x + 1)} \, dx$, write it in partial fraction form and then integrate each fraction separately as follows:

So $\displaystyle\int \dfrac{2x - 4}{(x - 3)(x + 1)} \, dx = \int \left(\dfrac{1}{2(x - 3)} + \dfrac{3}{2(x + 1)} \right) dx$

$= \displaystyle\int \left(\dfrac{1}{2}\left(\dfrac{1}{x - 3}\right) + \dfrac{3}{2}\left(\dfrac{1}{x + 1}\right) \right) dx$

$= \tfrac{1}{2} \ln|x - 3| + \tfrac{3}{2} \ln|x + 1| + c$

Tip:
Spot the use of
$\displaystyle\int \dfrac{f'(x)}{f(x)} \, dx = \ln|f(x)| + c$
(C3 Section 5.2).

Example 6.12 **a** Express $\dfrac{2x + 1}{x(x - 3)}$ in partial fractions.

  b Hence show by integrating that $\displaystyle\int_4^6 \dfrac{2x + 1}{x(x - 3)} \, dx = \tfrac{1}{3} \ln k$, where k is an integer to be found.

Step 1: Set out the partial fractions.

a Let $\dfrac{2x+1}{x(x-3)} \equiv \dfrac{A}{x} + \dfrac{B}{x-3}$

Step 2: Add the fractions.

$\dfrac{2x+1}{x(x-3)} \equiv \dfrac{A(x-3) + Bx}{x(x-3)}$

Step 3: Equate the numerators.

So $2x + 1 \equiv A(x-3) + Bx$

Step 4: Substitute appropriate values or compare coefficients.

Substituting $x = 0$,

$1 = A \times (-3) + B \times 0$

$\Rightarrow \quad A = -\dfrac{1}{3}$

Substituting $x = 3$,

$7 = A \times 0 + B \times 3$

$\Rightarrow \quad B = \dfrac{7}{3}$

Tip: Substitute values of x that will make one of the terms zero.

Tip: Comparing coefficients gives the following:
constant: $1 = -3A$
x terms: $2 = A + B$
Solve to find A and B.

Step 5: Write out the partial fractions.

So $\dfrac{2x+1}{x(x-3)} \equiv \dfrac{-\frac{1}{3}}{x} + \dfrac{\frac{7}{3}}{x-3}$

$\equiv -\dfrac{1}{3x} + \dfrac{7}{3(x-3)}$

Step 1: Integrate the partial fractions separately.

b $\displaystyle\int_4^6 \dfrac{2x+1}{x(x-3)}\, dx = \int_4^6 \left(-\dfrac{1}{3x} + \dfrac{7}{3(x-3)} \right) dx$

$= \displaystyle\int_4^6 \left\{ -\dfrac{1}{3}\left(\dfrac{1}{x}\right) + \dfrac{7}{3}\left(\dfrac{1}{x-3}\right) \right\} dx$

Tip: Take out numerical factors.

Tip: Recognise the log integrals (C3 Section 5.2).

$= \left[-\dfrac{1}{3}\ln|x| + \dfrac{7}{3}\ln|x-3| \right]_4^6$

Step 2: Substitute the limits.

$= -\dfrac{1}{3}\ln 6 + \dfrac{7}{3}\ln 3 - (-\dfrac{1}{3}\ln 4 + \dfrac{7}{3}\ln 1)$

Recall: $\ln 1 = 0$.

Step 3: Arrange into the required format.

$= -\dfrac{1}{3}\ln 6 + \dfrac{7}{3}\ln 3 + \dfrac{1}{3}\ln 4$

$= \dfrac{1}{3}(-\ln 6 + \ln 3^7 + \ln 4)$

$= \dfrac{1}{3}\ln \dfrac{4 \times 3^7}{6}$

$= \dfrac{1}{3}\ln(1458)$

Recall: Log laws
$n \log a = \log a^n$
$\log a - \log b = \log \dfrac{a}{b}$
(C2 Section 4.2).

Example 6.13

a Express $\dfrac{1}{x^2(x+1)}$ in partial fractions.

b Hence show that $\displaystyle\int_1^2 \dfrac{1}{x^2(x+1)}\, dx = \dfrac{1}{2} + \ln\dfrac{3}{4}$.

Step 1: Set out the partial fractions.

a Let $\dfrac{1}{x^2(x+1)} \equiv \dfrac{A}{x} + \dfrac{B}{x^2} + \dfrac{C}{x+1}$

Recall: Format when there are repeated linear factors in the denominator (Section 1.3).

Step 2: Add the fractions.

$\dfrac{1}{x^2(x+1)} \equiv \dfrac{Ax(x+1) + B(x+1) + Cx^2}{x^2(x+1)}$

Step 3: Equate the numerators.

So $1 \equiv Ax(x+1) + B(x+1) + Cx^2$

Tip: Expanding the brackets here could make your working more complicated.

Step 4: Substitute appropriate values or compare coefficients.

Substituting $x = 0$,

$1 = A \times 0 + B \times 1 + C \times 0$

$\Rightarrow \quad B = 1$

Substituting $x = -1$,
$$1 = A \times 0 + B \times 0 + C \times (-1)^2$$
$\Rightarrow \quad C = 1$

Equating coefficients of x^2,
$$0 = A + C$$
$\Rightarrow \quad A = -1$

Tip: Substitute $x = -1$ because then the factor $(x + 1)$ is equal to zero.

Step 5: Write out the partial fractions.
$$\frac{1}{x^2(x+1)} \equiv -\frac{1}{x} + \frac{1}{x^2} + \frac{1}{x+1}$$

Step 1: Integrate the partial fractions separately.

b $\displaystyle\int_1^2 \frac{1}{x^2(x+1)}\,dx = \int_1^2 \left(-\frac{1}{x} + \frac{1}{x^2} + \frac{1}{x+1}\right)dx$

Tip: Write x^n in index form (except when $n = -1$).

$$= \int_1^2 \left(-\frac{1}{x} + x^{-2} + \frac{1}{x+1}\right)dx$$

Step 2: Substitute the limits.

$$= \left[-\ln|x| - x^{-1} + \ln|x+1|\right]_1^2$$

Tip: Recognise the log integrals (C3 Section 5.2).

Step 3: Arrange into the required format.

$$= -\ln 2 - 2^{-1} + \ln 3 - (-\ln 1 - 1^{-1} + \ln 2)$$
$$= -\ln 2 - \tfrac{1}{2} + \ln 3 + 1 - \ln 2$$
$$= \tfrac{1}{2} + \ln 3 - 2\ln 2$$
$$= \tfrac{1}{2} + \ln 3 - \ln 4$$
$$= \tfrac{1}{2} + \ln \tfrac{3}{4}$$

Recall: Log laws
$n \log a = \log a^n$
$\log a - \log b = \log \dfrac{a}{b}$
(C2 Section 4.2).

SKILLS CHECK 6C: Integration using partial fractions

1 $f(x) \equiv \dfrac{x - 11}{(3x+1)(2x-5)} \equiv \dfrac{A}{3x+1} + \dfrac{B}{2x-5}$.

  a Find the values of the constants A and B.

  b Hence find $\displaystyle\int \frac{x-11}{(3x+1)(2x-5)}\,dx$.

  c Hence show that $\displaystyle\int_1^2 \frac{x-11}{(3x+1)(2x-5)}\,dx = \tfrac{2}{3}\ln\tfrac{7}{4} + \tfrac{1}{2}\ln 3$.

2 a Express $\dfrac{2x^2 - 9x - 31}{(x+2)(2x-1)(x+3)}$ in partial fractions.

  b Hence evaluate $\displaystyle\int_{-1}^0 \frac{2x^2 - 9x - 31}{(x+2)(2x-1)(x+3)}\,dx$, expressing your answer as a single natural logarithm.

3 a Given that $\dfrac{x^2}{x^2 - 9} \equiv A + \dfrac{B}{x+3} + \dfrac{C}{x-3}$, find the values of the constants A, B and C.

  b Hence find $\displaystyle\int \frac{x^2}{x^2 - 9}\,dx$.

4 a Express $\dfrac{1}{x(x-1)^2}$ in partial fractions.

b Hence find $\displaystyle\int_2^3 \dfrac{1}{x(x-1)^2}\,dx$ in the form $a + \ln b$ where a and b are rational numbers.

5 $f(x) \equiv \dfrac{16}{x^2(4-x)} = \dfrac{A}{x} + \dfrac{B}{x^2} + \dfrac{C}{4-x}$.

a Find the values of the constants A, B and C.

The diagram below shows the graph $y = \dfrac{4}{x\sqrt{4-x}}$, $x > 0$.

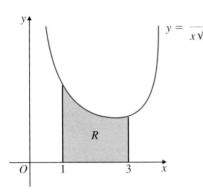

The region R, enclosed by the curve, the x-axis and the lines $x = 1$ and $x = 3$, is rotated through $360°$ about the x-axis.

Hint: Recall volumes of revolution (C3 Section 5.4).

b Find the volume of the solid generated, giving your answer to three significant figures.

SKILLS CHECK **6C EXTRA** is on the CD

Examination practice 6: Differentiation and integration

1 a Solve the differential equation $\dfrac{dy}{dx} = \dfrac{1}{y^2}$ giving the general solution for y in terms of x.

b Find the particular solution of this differential equation for which $y = -1$ when $x = 1$.

[AQA June 2002]

2 Given that $y = 0$ at $x = 1$, solve the differential equation $\dfrac{dy}{dx} = e^{x+y}$ giving your answer exactly in the form $y = f(x)$.

3 a Use integration by parts to find

$$\int x e^x \, dx$$

b Hence find the solution of the differential equation

$$\dfrac{dy}{dx} = yxe^x$$

given that $y = e$ when $x = 1$.

[AQA Jan 2003]

4 A group of students are researching the rate at which ice thickens on a frozen pond.
They have experimental evidence that when the air temperature is $-T\,°C$, the ice thickens at a rate

$$\frac{T}{14\,000x}\,\text{cm s}^{-1},$$

where x cm is the thickness of the ice that has already formed.

On a particular winter day the air temperature is constant at $-7\,°C$. At 12.00 noon the students note that the ice is 2 cm thick. Time t seconds later the thickness of the ice is x cm.

a Show that

$$\frac{dx}{dt} = \frac{1}{2000x}.$$

b Solve the differential equation and hence find the time when the students predict the ice will be 3 cm thick. [AQA Jan 2003]

5 Initially there are 2000 fish in a lake. The number of fish, x, at time t months later is modelled by the differential equation

$$\frac{dx}{dt} = x(1 - kt),$$

where k is a constant.

a Solve this differential equation to show that

$$x = 2000\,e^{t - \frac{1}{2}kt^2}.$$

b After 12 months the number of fish is again 2000. Find the value of k. [AQA June 2004]

6 The speed v m s^{-1} of a pebble falling through still water after t seconds can be modelled by the differential equation

$$\frac{dv}{dt} = 10 - 5v.$$

A pebble is placed carefully on the surface of the water at time $t = 0$ and begins to sink.

a Show that $t = \frac{1}{5}\ln\left(\frac{2}{2-v}\right)$.

b Use the model to find the speed of the pebble after 0.5 seconds, giving your answer to two significant figures. [AQA Jan 2004]

7 The rate of increase in population of bacteria is proportional to the size of the population that exists at any particular time.

a Explain briefly why this situation can be modelled by a differential equation of the form

$$\frac{dP}{dt} = kP,$$

where P is the size of the population, k is a constant and t is the time in minutes measured from a given starting time.

b i At time $t = 0$, the population of bacteria of type A is 1000. After 30 minutes, this population is 2000.
Solve the differential equation, stating the value of k in the form $q \ln 2$, where q is a rational number to be found.

ii As the population of type A increases, the population of another type of bacteria, B, decreases. The population, Q, of the type B bacteria at time t minutes is modelled by

$$Q = 5000\,e^{-0.05t}.$$

Find, to the nearest minute, the value of t when the populations of bacteria of types A and B are the same. [AQA June 2001]

8 Karen is designing a light show. She wants circles of light to appear on a big screen and to increase in size. In her design, a circle will initially have a radius of 50 cm, which will increase to 250 cm in 5 seconds. The radius of a circle is r cm at time t seconds. Karen considers two possible designs, A and B.

    a **Design A** The rate of increase of the radius is constant.

        Find the radius of the circle when $t = 2$.

    b **Design B** Karen wants the rate at which the radius increases to slow down as t increases.

        She bases this design on the model $\dfrac{dr}{dt} = \dfrac{k}{r}$, where k is a constant.

        i Solve this differential equation and show that $k = 6000$.

        ii Find the radius of the circle when $t = 2$.

        iii Show that, in Design B, the rate at which the area of the circle increases is constant.

        [AQA Jan 2005]

9 The gradient of a curve, C, at the point (x, y) is given by

$$\frac{dy}{dx} = \frac{1}{2y(x+2)}, \quad x > 0, \ y > 0$$

    The point $P(1, 1)$ lies on the curve C.

    a **i** Write down the gradient of the curve C at the point P.

        ii Show that the equation of the normal at P is $y + 6x = 7$.

    b Find the equation of the curve C in the form $y^2 = f(x)$.   [AQA June 2004]

10 A curve has equation $\dfrac{x^2}{9} + \dfrac{y^2}{25} = 1$.

    a Find the y-coordinates of the two points on the curve at which the x-coordinate is 2.

    b Find the values of the gradient of the curve at these two points, giving your answers to two significant figures.   [AQA June 2002]

11 A curve is given by the equation $9(y + 2)^2 = 5 + 4(x - 1)^2$.

    a Find the coordinates of the two points on the curve where $x = 2$.

    b Find the gradient of the curve at each of these points.   [AQA June 2004]

12 A curve has implicit equation

$$y^3 + xy = 4x - 2$$

    a Show that the value of $\dfrac{dy}{dx}$ at the point $(1, 1)$ is $\tfrac{3}{4}$.

    b Find the equation of the normal to the curve at the point $(1, 1)$.   [AQA Jan 2003]

13 A curve is given by the parametric equations

$$x = 1 - t^2, \quad y = 2t.$$

    a Find $\dfrac{dy}{dx}$ in terms of t.

    b Hence find the equation of the normal to the curve at the point where $t = 3$.   [AQA June 2002]

14 A curve is defined by the parametric equations

$$x = 3 \sin t \quad \text{and} \quad y = \cos t.$$

  a Show that, at the point P where $t = \dfrac{\pi}{4}$, the gradient of the curve is $-\dfrac{1}{3}$.

  b Find the equation of the tangent to the curve at the point P, giving your answer in the form $y = mx + c$. [AQA Jan 2003]

15 A curve is given by the parametric equations

$$x = 3t^2, \quad y = 6t.$$

  a i Find $\dfrac{dy}{dx}$ in terms of t.

  ii Find the gradient of the curve at the point where $t = \tfrac{1}{2}$.

  b i Find the equation of the curve in the form $x = f(y)$.

  ii Find $\dfrac{dx}{dy}$ in terms of y and hence verify your answer to part **a ii**. [AQA Jan 2004]

16 a Express $\dfrac{30}{(x + 4)(7 - 2x)}$ in the form $\dfrac{A}{x + 4} + \dfrac{B}{7 - 2x}$.

  b Hence find

$$\int_0^3 \dfrac{30}{(x + 4)(7 - 2x)} \, dx,$$

giving your answer in the form $p \ln q$, where p and q are rational numbers. [AQA June 2004]

17 a Express $\dfrac{13 - 2x}{(x + 4)(2x + 1)}$ in partial fractions.

  b Hence, prove that

$$\int_0^4 \dfrac{13 - 2x}{(x + 4)(2x + 1)} \, dx = p \ln 3 - q \ln 2$$

where p and q are positive integers. [AQA June 2004]

18 a i Show that

$$\dfrac{x^2}{x^2 - 16} = 1 + \dfrac{16}{x^2 - 16}.$$

  ii Express

$$\dfrac{16}{x^2 - 16} \text{ in the form } \dfrac{A}{x - 4} + \dfrac{B}{x + 4}.$$

  b Hence find

$$\int_5^8 \dfrac{x^2}{x^2 - 16} \, dx,$$

giving your answer in the form $p + q \ln r$. [AQA Jan 2003]

19 a Express $\dfrac{5x^2 - 8x + 1}{2x(x-1)^2}$ in the form $\dfrac{A}{x} + \dfrac{B}{x-1} + \dfrac{C}{(x-1)^2}$.

b Hence find $\displaystyle\int \dfrac{5x^2 - 8x + 1}{2x(x-1)^2}\,dx$.

c Hence show that $\displaystyle\int_4^9 \dfrac{5x^2 - 8x + 1}{2x(x-1)^2}\,dx = \ln\tfrac{32}{3} - \tfrac{5}{24}$.

20 The diagram shows part of a curve.

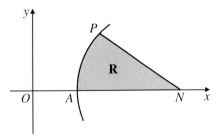

This curve is defined parametrically by
$$x = 4t + \dfrac{1}{t}, \quad y = 4t - \dfrac{1}{t}, \quad t > 0$$

The point P on the curve is where $t = 1$.
The normal to the curve at P intersects the x-axis at N.
The curve cuts the positive x-axis at the point A.

a Show that $t = \tfrac{1}{2}$ at the point A.

b Show that $\dfrac{dy}{dx} = \dfrac{4t^2 + 1}{4t^2 - 1}$.

c i Find an equation of the normal PN.
  ii Hence show that the x-coordinate of N is 10.

d i Express $x + y$ and $x - y$ in terms of t.
  ii Hence find a cartesian equation for the curve.

e The region **R**, bounded by the curve, the normal PN and the x-axis, is shown shaded in the diagram. Using your answer to part **d ii** and given that the area of **R** is $15 - 8\ln 2$, find the exact value of $\displaystyle\int_4^5 \sqrt{x^2 - 16}\,dx$.

[AQA Jan 2005]

7 Vectors

7.1 Vector geometry

Vectors in two and three dimensions. Magnitude of a vector. Algebraic operations of vector addition and multiplication by scalars and their geometrical interpretations.

A **scalar** is a quantity that can be expressed by magnitude (size) alone, for example length, distance, speed, volume.

A **vector** is a quantity that it is expressed in terms of magnitude and direction, for example displacement, velocity, acceleration, momentum.

So, as an example, wind speed is a scalar quantity and can be expressed in terms of its magnitude, such as 50 km/h. However, wind velocity is a vector quantity and therefore needs a direction given as well, for example 50 km/h from the south-west.

Vector notation

A directed line segment is drawn to represent a vector. The length of the line represents the magnitude of the vector and an arrow is used to represent the direction of the vector.

Note: A directed line segment is simply a straight line with an arrow to show direction.

For example, the vector of the displacement from P to Q can be represented as follows:

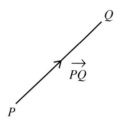

or, simply,

Note: In print, a lower case letter will be identified as a vector by the use of **bold** type. In your work you must use an underlined lower case letter, a̲.

where $\overrightarrow{PQ} = \mathbf{a}$

Vectors are **equal** if they have the same magnitude and the same direction.

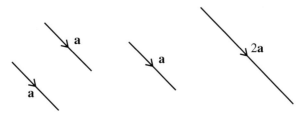

Note: Each of the three lines on the left of the diagram represents the same vector, **a**, because the lines are all of the same length and direction. The vector on the right, 2**a**, is a vector in the same direction as **a**, but with twice the magnitude.

Any vector **parallel** to a vector **a** can be written in the form $\lambda\mathbf{a}$, where λ is a scalar multiple. The vector $\lambda\mathbf{a}$ will have magnitude λ times the magnitude of **a**.

Recall: A scalar is a number.

Addition of vectors

To add vectors use the **triangle law** for vector addition.

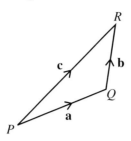

Tip:
Trace your finger along **a** then **b** in the direction of the arrows. You are starting at P and ending at R, which is the same outcome as tracing your finger along **c**.

Thinking of $\overrightarrow{PQ}$ as a displacement vector equivalent to travelling from P to Q, and similarly $\overrightarrow{QR}$ as a journey from Q to R, then $\overrightarrow{PQ} + \overrightarrow{QR}$ is the **resultant** journey from P to R, $\overrightarrow{PR}$, i.e.

$$\overrightarrow{PR} = \overrightarrow{PQ} + \overrightarrow{QR}$$
or $\mathbf{c} = \mathbf{a} + \mathbf{b}$

Subtraction of vectors

To work out $\mathbf{a} - \mathbf{b}$ think of it as $\mathbf{a} + (-\mathbf{b})$, so that you are adding the vector that is the same magnitude as **b** but is in the opposite direction.

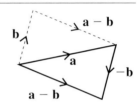

Tip:
Tracing your finger backwards along **b**, then along **a** means that you move the same distance and direction as tracing along **a** and then $-\mathbf{b}$.

Example 7.1 In the diagram, $\overrightarrow{AB} = \mathbf{p}$, $\overrightarrow{AF} = \mathbf{q}$ and $\overrightarrow{BF} = \overrightarrow{FC}$.

a Find, in terms of **p** and **q**,
  i $\overrightarrow{BF}$
  ii $\overrightarrow{BC}$
  iii $\overrightarrow{AC}$

b D and E are the midpoints of AB and AF.
  i Find $\overrightarrow{DE}$ in terms of **p** and **q**.
  ii Prove that DE is parallel to BF.

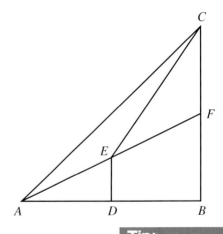

Step 1: Use the triangle law to add vectors.

a i $\overrightarrow{BF} = \overrightarrow{BA} + \overrightarrow{AF}$
        $= -\mathbf{p} + \mathbf{q}$
        $= \mathbf{q} - \mathbf{p}$

Tip:
Use $\triangle ABF$: $\overrightarrow{BA} = -\overrightarrow{AB} = -\mathbf{p}$

Step 2: Use the given relation with your answer from **a**.

  ii $\overrightarrow{BF} = \overrightarrow{FC}$
      so $\overrightarrow{BC} = 2\overrightarrow{BF}$
      $\overrightarrow{BC} = 2(\mathbf{q} - \mathbf{p})$

Step 3: Use the triangle law to add vectors.

  iii $\overrightarrow{AC} = \overrightarrow{AB} + \overrightarrow{BC}$
       $= \mathbf{p} + 2(\mathbf{q} - \mathbf{p})$
       $= \mathbf{p} + 2\mathbf{q} - 2\mathbf{p} = 2\mathbf{q} - \mathbf{p}$

Tip:
Use $\triangle ABC$.

Step 1: Use the given relations and the triangle law to add vectors.

b i $\overrightarrow{AD} = \frac{1}{2}\overrightarrow{AB} = \frac{1}{2}\mathbf{p}$
$\overrightarrow{AE} = \frac{1}{2}\overrightarrow{AF} = \frac{1}{2}\mathbf{q}$
$\overrightarrow{DE} = \overrightarrow{DA} + \overrightarrow{AE}$
$= -\frac{1}{2}\mathbf{p} + \frac{1}{2}\mathbf{q}$
$= \frac{1}{2}\mathbf{q} - \frac{1}{2}\mathbf{p}$

Tip:
In **b i**, if D is the midpoint of AB, then the vector $\overrightarrow{AD}$ must be in the same direction as $\overrightarrow{AB}$ and half its magnitude.

Tip:
Use $\triangle ADE$: $\overrightarrow{DA} = -\overrightarrow{AD} = -\frac{1}{2}\mathbf{p}$

Step 2: Write $\overrightarrow{DE}$ in terms of $\overrightarrow{BF}$ and state your conclusion.

ii $\overrightarrow{DE} = \frac{1}{2}\mathbf{q} - \frac{1}{2}\mathbf{p} = \frac{1}{2}(\mathbf{q} - \mathbf{p}) = \frac{1}{2}\overrightarrow{BF}$

Hence DE is parallel to BF.

Recall:
Two vectors are parallel if one is a scalar multiple of the other.

Note:
This proves an important theorem in geometry called the midpoint theorem for triangles.

Magnitude of a vector in two dimensions

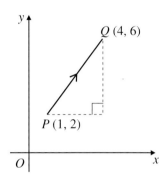

Note:
To get from P to Q you go 3 units right and 4 units up.

If P is the point $(1, 2)$ and Q is the point $(4, 6)$, then

$\overrightarrow{PQ} = \begin{bmatrix} 3 \\ 4 \end{bmatrix}$

where 3 is the increase in the x-coordinate and 4 is the increase in the y-coordinate.

The vector with the same magnitude as $\overrightarrow{PQ}$, but in the opposite direction, is $\overrightarrow{QP}$, where

$\overrightarrow{QP} = \begin{bmatrix} -3 \\ -4 \end{bmatrix}$

So $\overrightarrow{QP} = -\overrightarrow{PQ}$.

The **magnitude** of the vector $\overrightarrow{PQ}$ is also called its **modulus** and is written PQ or $|\overrightarrow{PQ}|$. This can be found using Pythagoras' theorem, where

$PQ = |\overrightarrow{PQ}| = \sqrt{3^2 + 4^2} = 5$

In general, if $\mathbf{v} = \begin{bmatrix} x \\ y \end{bmatrix}$, then $|\mathbf{v}| = \sqrt{x^2 + y^2}$.

Tip:
Vectors are written as a column. Don't confuse them with coordinates which are written in a row.

Tip:
To get from Q to P you go 3 units left and 4 units down.

Note:
This is true for all vectors.

Tip:
If you notice it is a 3, 4, 5 triangle you could use this fact to save having to work it out.

Note:
If you are using lower case letters for the vector, e.g. $\mathbf{v}$, then the magnitude is written $|\mathbf{v}|$.

65

Magnitude of a vector in three dimensions

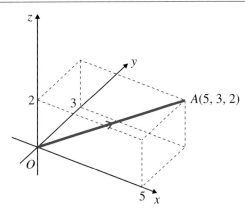

Note:
If O is the origin, then $\overrightarrow{OA}$ is a **position vector** (see Section 7.2).

Tip:
Imagine the x- and y-axes flat on the table, then the z-axis rises perpendicularly upwards.

The vector that joins the points $O(0, 0, 0)$ and $A(5, 3, 2)$ is $\overrightarrow{OA}$,

where $\overrightarrow{OA} = \begin{bmatrix} 5 \\ 3 \\ 2 \end{bmatrix}$.

As before, to find the magnitude or modulus of a vector, use Pythagoras' theorem in three dimensions, so $|\overrightarrow{OA}| = \sqrt{5^2 + 3^2 + 2^2} = \sqrt{38}$.

Tip:
Leave your value in surd form where necessary.

In general, if $\mathbf{v} = \begin{bmatrix} x \\ y \\ z \end{bmatrix}$, then $|\mathbf{v}| = \sqrt{x^2 + y^2 + z^2}$.

Unit vectors

A **unit vector** has magnitude 1.

To find a unit vector in the direction of the vector $\mathbf{v}$, divide $\mathbf{v}$ by its magnitude, so the unit vector is $\dfrac{\mathbf{v}}{|\mathbf{v}|}$.

For the vector $\overrightarrow{PQ}$ given above, a unit vector parallel to $\overrightarrow{PQ}$ is

$$\dfrac{\overrightarrow{PQ}}{|\overrightarrow{PQ}|} = \tfrac{1}{5}\overrightarrow{PQ} = \begin{bmatrix} \frac{3}{5} \\ \frac{4}{5} \end{bmatrix}$$

For the vector $\overrightarrow{OA}$ given above, a unit vector parallel to $\overrightarrow{OA}$ is

$$\dfrac{\overrightarrow{OA}}{|\overrightarrow{OA}|} = \tfrac{1}{\sqrt{38}}\overrightarrow{OA} = \begin{bmatrix} \frac{5}{\sqrt{38}} \\ \frac{3}{\sqrt{38}} \\ \frac{2}{\sqrt{38}} \end{bmatrix}$$

Example 7.2 Given the vectors $\mathbf{p} = \begin{bmatrix} 4 \\ 2 \\ -5 \end{bmatrix}$ and $\mathbf{q} = \begin{bmatrix} 2 \\ -1 \\ 2 \end{bmatrix}$, find

a $\mathbf{p} + 2\mathbf{q}$,

b i $\mathbf{p} - \mathbf{q}$,
 ii the exact value of $|\mathbf{p} - \mathbf{q}|$,

c a unit vector in the direction of $\mathbf{p}$.

Step 1: Add the vectors. **a** $\mathbf{p} + 2\mathbf{q} = \begin{bmatrix} 4 \\ 2 \\ -5 \end{bmatrix} + 2\begin{bmatrix} 2 \\ -1 \\ 2 \end{bmatrix} = \begin{bmatrix} 4 \\ 2 \\ -5 \end{bmatrix} + \begin{bmatrix} 4 \\ -2 \\ 4 \end{bmatrix} = \begin{bmatrix} 8 \\ 0 \\ -1 \end{bmatrix}$

Step 1: Find $\mathbf{p} - \mathbf{q}$. **b i** $\mathbf{p} - \mathbf{q} = \begin{bmatrix} 4 \\ 2 \\ -5 \end{bmatrix} - \begin{bmatrix} 2 \\ -1 \\ 2 \end{bmatrix} = \begin{bmatrix} 2 \\ 3 \\ -7 \end{bmatrix}$

Note:
$|\mathbf{p} - \mathbf{q}| \ne |\mathbf{p}| - |\mathbf{q}|$

Step 2: Use Pythagoras' theorem to find the modulus. **ii** $|\mathbf{p} - \mathbf{q}| = \sqrt{2^2 + 3^2 + (-7)^2} = \sqrt{62}$

Tip:
You are asked for the exact value, so leave your answer in surd form.

Step 1: Use Pythagoras' theorem to find the modulus. **c** $\mathbf{p} = \begin{bmatrix} 4 \\ 2 \\ -5 \end{bmatrix}$

so $|\mathbf{p}| = \sqrt{4^2 + 2^2 + (-5)^2} = \sqrt{45} = 3\sqrt{5}$

Note:
It is good practice to simplify surds.

Step 2: Divide the vector by its modulus. Unit vector is $\dfrac{\mathbf{p}}{|\mathbf{p}|} = \dfrac{1}{3\sqrt{5}} \begin{bmatrix} 4 \\ 2 \\ -5 \end{bmatrix} = \begin{bmatrix} \frac{4}{3\sqrt{5}} \\ \frac{2}{3\sqrt{5}} \\ -\frac{5}{3\sqrt{5}} \end{bmatrix}$

Note:
You could write this as $\begin{bmatrix} \frac{4\sqrt{5}}{15} \\ \frac{2\sqrt{5}}{15} \\ -\frac{\sqrt{5}}{3} \end{bmatrix}$.

Example 7.3 Given the vector $\mathbf{p} = \begin{bmatrix} 3 \\ -1 \\ 2 \end{bmatrix}$, state whether each of the following vectors is equal to **p**, parallel to **p**, or neither.

a $\begin{bmatrix} -6 \\ 2 \\ -4 \end{bmatrix}$ **b** $\begin{bmatrix} 6 \\ -2 \\ -4 \end{bmatrix}$ **c** $-\frac{1}{3}\begin{bmatrix} 9 \\ 3 \\ -6 \end{bmatrix}$

Step 1: Check whether the vector is a scalar multiple of **p** or equal to **p**. **a** $\begin{bmatrix} -6 \\ 2 \\ -4 \end{bmatrix} = -2\begin{bmatrix} 3 \\ -1 \\ 2 \end{bmatrix}$, so the given vector is parallel to **p**.

Tip:
The scalar multiple is -2.

b $\begin{bmatrix} 6 \\ -2 \\ -4 \end{bmatrix} = 2\begin{bmatrix} 3 \\ -1 \\ -2 \end{bmatrix}$, so it is not a multiple of **p**.

Hence the given vector is neither parallel nor equal to **p**.

c $-\frac{1}{3}\begin{bmatrix} 9 \\ 3 \\ -6 \end{bmatrix} = \begin{bmatrix} 3 \\ -1 \\ 2 \end{bmatrix}$, so the given vector is equal to **p**.

Example 7.4 Given the vectors $\overrightarrow{PQ} = \begin{bmatrix} 4 \\ 1 \\ -2 \end{bmatrix}$ and $\overrightarrow{RQ} = \begin{bmatrix} 2 \\ 1 \\ -1 \end{bmatrix}$, find $\overrightarrow{PR}$.

Step 1: Draw a sketch to represent the vectors.

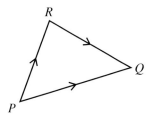

Note:
The diagram is not meant to be an accurate representation, just a sketch to help.

Step 2: Use the triangle law to add the vectors.
$\overrightarrow{PR} = \overrightarrow{PQ} + \overrightarrow{QR} = \overrightarrow{PQ} - \overrightarrow{RQ} = \begin{bmatrix} 4 \\ 1 \\ -2 \end{bmatrix} - \begin{bmatrix} 2 \\ 1 \\ -1 \end{bmatrix} = \begin{bmatrix} 2 \\ 0 \\ -1 \end{bmatrix}$

Recall:
$\overrightarrow{QR} = -\overrightarrow{RQ}$.

SKILLS CHECK 7A: Vector geometry

1 Given the vector $\mathbf{a} = \begin{bmatrix} 2 \\ -1 \\ 1 \end{bmatrix}$, state whether each of the following vectors is equal to $\mathbf{a}$, parallel to $\mathbf{a}$, or neither.

  a $\begin{bmatrix} -4 \\ 2 \\ 2 \end{bmatrix}$
  b $\frac{1}{2}\begin{bmatrix} 4 \\ -2 \\ 2 \end{bmatrix}$
  c $\begin{bmatrix} -8 \\ 4 \\ -4 \end{bmatrix}$

2 Given the vectors $\overrightarrow{AC} = \begin{bmatrix} 3 \\ 1 \\ -1 \end{bmatrix}$ and $\overrightarrow{BC} = \begin{bmatrix} -1 \\ -4 \\ 5 \end{bmatrix}$, find $\overrightarrow{AB}$.

3 Find the magnitude of each of the following vectors.

  a $\begin{bmatrix} 5 \\ 12 \end{bmatrix}$
  b $\begin{bmatrix} 2 \\ -3 \\ 1 \end{bmatrix}$
  c $\begin{bmatrix} -3 \\ 0 \\ 4 \end{bmatrix}$

  d $\begin{bmatrix} 2 \\ 4 \end{bmatrix}$
  e $\begin{bmatrix} 5 \\ -1 \\ 2 \end{bmatrix}$
  f $\begin{bmatrix} 1 \\ -2 \\ -1 \end{bmatrix}$

4 Given that $\mathbf{a} = \begin{bmatrix} 1 \\ -3 \\ 1 \end{bmatrix}$, $\mathbf{b} = \begin{bmatrix} 3 \\ -1 \\ 4 \end{bmatrix}$ and $\mathbf{c} = \begin{bmatrix} 2 \\ 1 \\ -1 \end{bmatrix}$, find

  a $\mathbf{a} + 2\mathbf{b}$
  b $2\mathbf{c} - \mathbf{a}$
  c $\mathbf{a} - \mathbf{b} + 3\mathbf{c}$
  d $|\mathbf{b} - \mathbf{a}|$
  e $|\mathbf{a} + \mathbf{b} + \mathbf{c}|$
  f $|2\mathbf{a} - \mathbf{b} + 2\mathbf{c}|$

5 Given that $\mathbf{a} = \begin{bmatrix} 2 \\ 0 \\ -1 \end{bmatrix}$, $\mathbf{b} = \begin{bmatrix} -1 \\ -2 \\ -1 \end{bmatrix}$ and $\mathbf{c} = \begin{bmatrix} 3 \\ 3 \\ 3 \end{bmatrix}$, find

  a $\mathbf{b} - 2\mathbf{c}$
  b $\mathbf{a} - \mathbf{b} + \mathbf{c}$
  c $3\mathbf{a} - \mathbf{b}$
  d $|\mathbf{a} + \mathbf{c}|$
  e $|2\mathbf{c} - \mathbf{a} - \mathbf{b}|$
  f $|2\mathbf{a} + 2\mathbf{b} - \mathbf{c}|$

6 Find a unit vector in the direction of $\begin{bmatrix} 1 \\ 5 \\ -7 \end{bmatrix}$.

7 Given the points $A(-1, 1)$, $B(1, -2)$ and $C(2, 3)$,

   a find the vectors

     i $\overrightarrow{AB}$ **ii** $\overrightarrow{AC}$ **iii** $\overrightarrow{BC}$

   b Find the magnitude of each of the vectors in part **a**.

   c Hence prove that ABC is a right-angled triangle.

8 Vectors **a** and **b** are such that $\mathbf{a} = \begin{bmatrix} 3 \\ 2 \\ 1 \end{bmatrix}$, $\mathbf{b} = \begin{bmatrix} -1 \\ 6 \\ \lambda \end{bmatrix}$ and $|\mathbf{a} - \mathbf{b}| = 6$.

Find the two possible values of λ.

SKILLS CHECK **7A EXTRA** is on the **CD**

7.2 Position vectors and distance

Position vectors. The distance between two points.

If you have a fixed origin, O, and a point, A, then the vector $\overrightarrow{OA} = \mathbf{a}$ is called the **position vector** of A.

So, for example, consider the point $A(2, 1, -1)$. A has position vector

$$\mathbf{a} = \overrightarrow{OA} = \begin{bmatrix} 2 \\ 1 \\ -1 \end{bmatrix}$$

Tip:
As before, be careful not to confuse the coordinates of A (written horizontally) with the column vector, which represents the displacement from the origin to A.

If the two points A and B have position vectors $\overrightarrow{OA} = \mathbf{a}$ and $\overrightarrow{OB} = \mathbf{b}$ respectively, then, using the triangle law,

$$\overrightarrow{AB} = \overrightarrow{AO} + \overrightarrow{OB}$$
$$= -\overrightarrow{OA} + \overrightarrow{OB}$$
$$= -\mathbf{a} + \mathbf{b}$$
$$= \mathbf{b} - \mathbf{a}$$

Recall:
Triangle law for vector addition (Section 6.1).

Example 7.5 Given the points A and B with position vectors **a** and **b** respectively, find the position vector of the midpoint of the line AB.

Let M be the midpoint of AB.

Step 1: Draw a diagram.

Step 2: Use the triangle law to add vectors.

Step 3: Simplify.

$$\overrightarrow{OM} = \overrightarrow{OA} + \overrightarrow{AM}$$
$$= \overrightarrow{OA} + \tfrac{1}{2}\overrightarrow{AB}$$
$$= \mathbf{a} + \tfrac{1}{2}(\mathbf{b} - \mathbf{a})$$
$$= \mathbf{a} + \tfrac{1}{2}\mathbf{b} - \tfrac{1}{2}\mathbf{a}$$
$$= \tfrac{1}{2}\mathbf{a} + \tfrac{1}{2}\mathbf{b}$$
$$= \tfrac{1}{2}(\mathbf{a} + \mathbf{b})$$

Tip:
If M is the midpoint of AB, then the vector $\overrightarrow{AM}$ is half the vector $\overrightarrow{AB}$.

Tip:
Use the result $\overrightarrow{AB} = \mathbf{b} - \mathbf{a}$, found earlier.

Note:
The position vector of M is $\overrightarrow{OM} = \tfrac{1}{2}(\mathbf{a} + \mathbf{b})$. This is different from the vector $\overrightarrow{AM}$.

The distance between two points

Consider the points $A(x_1, y_1, z_1)$ and $B(x_2, y_2, z_2)$.

Then $\mathbf{a} = \begin{bmatrix} x_1 \\ y_1 \\ z_1 \end{bmatrix}$ and $\mathbf{b} = \begin{bmatrix} x_2 \\ y_2 \\ z_2 \end{bmatrix}$.

$$\overrightarrow{AB} = \mathbf{b} - \mathbf{a} = \begin{bmatrix} x_2 \\ y_2 \\ z_2 \end{bmatrix} - \begin{bmatrix} x_1 \\ y_1 \\ z_1 \end{bmatrix} = \begin{bmatrix} x_2 - x_1 \\ y_2 - y_1 \\ z_2 - z_1 \end{bmatrix}$$

The distance, d, between A and B is given by

$$d = |\overrightarrow{AB}| = \sqrt{(x_2 - x_1)^2 + (y_2 - y_1)^2 + (z_2 - z_1)^2}$$

Recall: Pythagoras' theorem in three dimensions.

Example 7.6 Find the distance between the points $A(2, 4, 1)$ and $B(5, -2, 3)$.

Step 1: Find $\overrightarrow{AB}$. $\mathbf{a} = \begin{bmatrix} 2 \\ 4 \\ 1 \end{bmatrix}$, $\mathbf{b} = \begin{bmatrix} 5 \\ -2 \\ 3 \end{bmatrix}$ so $\overrightarrow{AB} = \mathbf{b} - \mathbf{a} = \begin{bmatrix} 5 \\ -2 \\ 3 \end{bmatrix} - \begin{bmatrix} 2 \\ 4 \\ 1 \end{bmatrix} = \begin{bmatrix} 3 \\ -6 \\ 2 \end{bmatrix}$

Step 2: Calculate the magnitude of the vector.
$|\overrightarrow{AB}| = \sqrt{3^2 + (-6)^2 + 2^2}$
$= \sqrt{49}$
$= 7$

The distance between A and B is 7.

SKILLS CHECK 7B: Position vectors and distance

1 Find the vector $\overrightarrow{AB}$ given that the position vectors of A and B relative to O are as follows.

a $\overrightarrow{OA} = \begin{bmatrix} 6 \\ 2 \\ -1 \end{bmatrix}$, $\overrightarrow{OB} = \begin{bmatrix} 2 \\ -3 \\ 1 \end{bmatrix}$ **b** $\mathbf{a} = \begin{bmatrix} 1 \\ 1 \\ -4 \end{bmatrix}$, $\mathbf{b} = \begin{bmatrix} 3 \\ -3 \\ 2 \end{bmatrix}$ **c** $\mathbf{a} = \begin{bmatrix} 3 \\ -2 \\ 1 \end{bmatrix}$, $\mathbf{b} = \begin{bmatrix} 0 \\ 3 \\ -1 \end{bmatrix}$

2 The points P, Q, and R have position vectors $2\mathbf{a} + \mathbf{b}$, $\mathbf{a} + 3\mathbf{b}$ and $-2\mathbf{a} + k\mathbf{b}$ respectively.

a Find **i** $\overrightarrow{PQ}$, **ii** $\overrightarrow{QR}$.

b Given that P, Q and R lie along a straight line, find the value of k.

c State the ratio $PQ : QR$.

3 Find the length of the line joining each of the following pairs of points, giving your answer to one decimal place.

a $A(4, -3, 7)$, $B(-2, -4, 3)$

b $A(-6, 6, 0)$, $B(2, 2, -2)$

c $A(1, 4, -6)$, $B(0, -9, 5)$

4 The points $A(4, 2, 3)$, $B(3, 3, -1)$, $C(6, 0, -1)$ and D form a parallelogram.

  a Show that $|\overrightarrow{AB}| = |\overrightarrow{BC}|$.

  b Find the position vector of the point D.

5 Given that the distance between the points $A(3, -2, 1)$ and $B(2, -4, p)$ is 3, find the two possible values of p.

SKILLS CHECK **7B EXTRA is on the CD**

7.3 Vector equations of lines

Vector equations of lines.

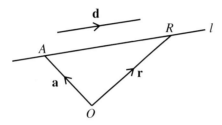

The diagram shows a line l, parallel to a vector **d**.
The point A, with position vector **a**, lies on the line.

Consider a general point $R(x, y, z)$ on the line, with position vector

$$\mathbf{r} = \begin{bmatrix} x \\ y \\ z \end{bmatrix}.$$

Since the line is parallel to **d**, $\overrightarrow{AR} = t\mathbf{d}$, where t is a scalar.

Now $\mathbf{r} = \overrightarrow{OR}$
$= \overrightarrow{OA} + \overrightarrow{AR}$
$= \mathbf{a} + t\mathbf{d}$

Hence, a vector equation of a straight line passing through a point A and parallel to a vector **d** is

$$\mathbf{r} = \mathbf{a} + t\mathbf{d}$$

Different values of the scalar t will give the position vectors of different points that lie on the line.

Note:
The vector **d** is called the **direction vector** of the line.

Recall:
Any vector parallel to **d** is a scalar multiple of **d**.

Note:
There are an infinite number of equations for a particular line: any point on the line could be used in place of A and any multiple of **d** could be used as the direction vector.

Note:
t is a parameter.
Other letters may be used, for example s, λ, μ, m, p.

Example 7.7 **a** Find a vector equation of the line l that passes through the point $A(3, -1, 2)$ and is parallel to the vector $\begin{bmatrix} 4 \\ 0 \\ -1 \end{bmatrix}$.

  b Find the coordinates of the point B on l, with parameter $t = 4$.

  c Show that the point P, with position vector $\begin{bmatrix} 11 \\ -1 \\ 0 \end{bmatrix}$, lies on l.

Step 1: State the position vector of a known point on the line and also the direction vector of the line.

a Since A lies on the line, $\mathbf{a} = \begin{bmatrix} 3 \\ -1 \\ 2 \end{bmatrix}$.

The direction vector $\mathbf{d} = \begin{bmatrix} 4 \\ 0 \\ -1 \end{bmatrix}$.

Step 2: Substitute into the vector equation of a line.

A vector equation of the line is $\mathbf{r} = \begin{bmatrix} 3 \\ -1 \\ 2 \end{bmatrix} + t \begin{bmatrix} 4 \\ 0 \\ -1 \end{bmatrix}$.

Step 1: Substitute $t = 4$ into the vector equation of the line.

b For all points on the line, $\mathbf{r} = \begin{bmatrix} 3 + 4t \\ -1 \\ 2 - t \end{bmatrix}$.

At the point B, $t = 4$, so $\mathbf{r} = \begin{bmatrix} 19 \\ -1 \\ -2 \end{bmatrix}$, i.e. $\begin{bmatrix} x \\ y \\ z \end{bmatrix} = \begin{bmatrix} 19 \\ -1 \\ -2 \end{bmatrix}$.

Recall:
A point with position vector $\begin{bmatrix} x \\ y \\ z \end{bmatrix}$ has coordinates (x, y, z).

Step 2: State the coordinates.

The point B has coordinates $(19, -1, -2)$.

Step 1: Substitute the position vector into the equation of the line.

c If P lies on the line then there is a value of t which satisfies

$$\begin{bmatrix} 3 + 4t \\ -1 \\ 2 - t \end{bmatrix} = \begin{bmatrix} 11 \\ -1 \\ 0 \end{bmatrix}$$

Step 2: Equate the vectors and attempt to solve for t.

i.e. $3 + 4t = 11 \Rightarrow t = 2$
$-1 = -1$ (no further information)
$2 - t = 0 \Rightarrow t = 2$

Since $t = 2$ satisfies all three equations, the point P lies on the line.

Consider now the case when all you know is two points, A and B, on the line. Let the position vectors of these points be $\mathbf{a}$ and $\mathbf{b}$.

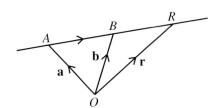

Since the line passes through A and B, the direction vector of the line is given by $\mathbf{d} = \overrightarrow{AB} = \mathbf{b} - \mathbf{a}$.

So $\mathbf{r} = \overrightarrow{OR}$
$= \overrightarrow{OA} + \overrightarrow{AR}$
$= \mathbf{a} + t(\mathbf{b} - \mathbf{a})$

Hence, a vector equation of a line through points A and B is

$$\mathbf{r} = \mathbf{a} + t(\mathbf{b} - \mathbf{a})$$

Note:
The vector $\overrightarrow{AR}$ is a scalar multiple of the vector $\overrightarrow{AB}$.

Note:
You could also write
$\mathbf{r} = \overrightarrow{OB} + \overrightarrow{BR} = \mathbf{b} + t(\mathbf{b} - \mathbf{a})$
for the equation of the line.

Example 7.8 **a** Find a vector equation of the straight line which passes through the points $A(0, -3, -3)$ and $B(2, 3, -1)$.

b Given that the point $P(p, 0, q)$ lies on the line, find p and q.

Step 1: Write down the position vectors of the given points.

a The position vector of A is $\mathbf{a} = \begin{bmatrix} 0 \\ -3 \\ -3 \end{bmatrix}$.

The position vector of B is $\mathbf{b} = \begin{bmatrix} 2 \\ 3 \\ -1 \end{bmatrix}$.

Recall:
If $\overrightarrow{OA} = \mathbf{a}$ and $\overrightarrow{OB} = \mathbf{b}$, then $\overrightarrow{AB} = \mathbf{b} - \mathbf{a}$ (Section 7.2)

Step 2: Find a direction vector

$\overrightarrow{AB} = \mathbf{b} - \mathbf{a} = \begin{bmatrix} 2 \\ 3 \\ -1 \end{bmatrix} - \begin{bmatrix} 0 \\ -3 \\ -3 \end{bmatrix} = \begin{bmatrix} 2 \\ 6 \\ 2 \end{bmatrix}$

Tip:
Use either of the given points to form the equation.

Step 3: Use an appropriate formula for the vector equation of a line.

So an equation of the line is $\mathbf{r} = \begin{bmatrix} 0 \\ -3 \\ -3 \end{bmatrix} + t \begin{bmatrix} 2 \\ 6 \\ 2 \end{bmatrix}$.

Tip:
This equation can also be written
$\mathbf{r} = \begin{bmatrix} 2t \\ -3 + 6t \\ -3 + 2t \end{bmatrix}$.

Step 1: Write down the position vector of P.

b $\overrightarrow{OP} = \begin{bmatrix} p \\ 0 \\ q \end{bmatrix}$

Step 2: Set up three simultaneous equations by equating coefficients.

If the point P lies on the line then

$\begin{bmatrix} 2t \\ -3 + 6t \\ -3 + 2t \end{bmatrix} = \begin{bmatrix} p \\ 0 \\ q \end{bmatrix}$.

$p = 2t$ ①
$0 = -3 + 6t$ ②
$q = -3 + 2t$ ③

Step 3: Solve an equation to find t.

From ②,
$0 = -3 + 6t$
$t = \frac{1}{2}$

Tip:
Start with the equation that has only one unknown.

Step 4: Substitute for t into the other equations to find p and q.

From ①,
$p = 2 \times \frac{1}{2} = 1$

From ③,
$q = -3 + 2 \times \frac{1}{2} = -2$

So $p = 1$ and $q = -2$.

Note:
The coordinates of P are $(1, 0, -2)$.

Intersection of lines

In two dimensions two distinct lines are either parallel or they intersect.
In three dimensions a pair of distinct lines may be parallel or intersecting, but they may be neither: such lines are called **skew**.

Note:
Skew lines are not parallel and they do not intersect.

Example 7.9 Line l_1 has vector equation $\mathbf{r} = \begin{bmatrix} 1 \\ 0 \\ 4 \end{bmatrix} + s\begin{bmatrix} -1 \\ 3 \\ 2 \end{bmatrix}$ and line l_2 has vector equation $\mathbf{r} = \begin{bmatrix} 0 \\ -4 \\ 3 \end{bmatrix} + t\begin{bmatrix} 2 \\ 1 \\ -1 \end{bmatrix}$.

Note: Since there are two lines in this question a different parameter is used in each equation.

Show that l_1 and l_2 intersect, and find the coordinates of the point of intersection.

Step 1: Rewrite the equations in an appropriate format.

The lines are $\mathbf{r} = \begin{bmatrix} 1-s \\ 3s \\ 4+2s \end{bmatrix}$ and $\mathbf{r} = \begin{bmatrix} 2t \\ -4+t \\ 3-t \end{bmatrix}$.

Step 2: Set up three simultaneous equations by equating coefficients.

At any point of intersection,

$\begin{bmatrix} 1-s \\ 3s \\ 4+2s \end{bmatrix} = \begin{bmatrix} 2t \\ -4+t \\ 3-t \end{bmatrix}$

$\Rightarrow \quad 1 - s = 2t \qquad ①$
$\quad\quad\quad 3s = -4 + t \qquad ②$
$\quad\quad\quad 4 + 2s = 3 - t \qquad ③$

Note: There are *three* equations connecting *two* unknowns. Use two of the equations to find values for s and t, then check for compatibility with the third equation.

Step 3: Solve two of the equations to find s and t.

Adding ② and ③,

$4 + 5s = -1$
$5s = -5$
$s = -1$

Substituting in ②,

$3(-1) = -4 + t$
$t = 1$

Note: If the values do not satisfy the third equation, then there are no solutions. In this case, the lines do not intersect. Provided they are not parallel, they are skew.

Step 4: Check whether the third equation is satisfied and state a conclusion.

Checking in ①,

LHS $= 1 - (-1) = 2$
RHS $= 2 \times 1 = 2$

Since LHS = RHS, equation ① is satisfied.
Therefore the lines intersect.

Tip: You must show that you have checked in the third equation or you will lose marks.

Tip: As an extra check substitute s and t into all three equations.

Step 5: Substitute one of the parameter values into the corresponding vector equation.

At the point of intersection, $s = -1$ and $t = 1$.

Substituting $s = -1$ into the equation of l_1 gives $\mathbf{r} = \begin{bmatrix} 2 \\ -3 \\ 2 \end{bmatrix}$.

So the coordinates of the point of intersection are $(2, -3, 2)$.

Tip: You could substitute $t = 1$ into the equation for l_2 instead.

Tip: Remember to write your answer as coordinates as requested.

SKILLS CHECK **7C: Vector equations of lines**

1 Find a vector equation of the straight line which passes through the point A, with position vector $\mathbf{a}$, and is parallel to the vector $\mathbf{d}$.

a $\mathbf{a} = \begin{bmatrix} 2 \\ 0 \\ -1 \end{bmatrix}, \mathbf{d} = \begin{bmatrix} 7 \\ -2 \\ 6 \end{bmatrix}$ **b** $\mathbf{a} = \begin{bmatrix} 4 \\ -1 \\ 3 \end{bmatrix}, \mathbf{d} = \begin{bmatrix} 0 \\ 6 \\ -1 \end{bmatrix}$ **c** $\mathbf{a} = \begin{bmatrix} -1 \\ 2 \\ 1 \end{bmatrix}, \mathbf{d} = \begin{bmatrix} -1 \\ -1 \\ 0 \end{bmatrix}$

2 Find a vector equation of the straight line which passes through each of the following pairs of points.

    a $A(2, -1, 5), B(-3, 0, 1)$ **b** $A(0, 2, 1), B(3, 3, -1)$ **c** $A(1, 4, -2), B(-3, 1, 4)$

3 Show that the point $P(2, -1, 7)$ lies on the line $\mathbf{r} = \begin{bmatrix} 4 \\ -2 \\ 3 \end{bmatrix} + t \begin{bmatrix} 2 \\ -1 \\ -4 \end{bmatrix}$.

4 Given that the point $(a, b, 0)$ lies on the line $\mathbf{r} = \begin{bmatrix} 6 \\ -4 \\ 2 \end{bmatrix} + \lambda \begin{bmatrix} 1 \\ -1 \\ 1 \end{bmatrix}$, find a and b.

5 Prove that the following pairs of lines are skew.

    a $\mathbf{r} = \begin{bmatrix} 4 \\ 0 \\ 9 \end{bmatrix} + s \begin{bmatrix} 1 \\ 2 \\ 5 \end{bmatrix}$ and $\mathbf{r} = \begin{bmatrix} -7 \\ 3 \\ -1 \end{bmatrix} + t \begin{bmatrix} 2 \\ -1 \\ 3 \end{bmatrix}$

    b $\mathbf{r} = \begin{bmatrix} 1 \\ -8 \\ 19 \end{bmatrix} + \lambda \begin{bmatrix} -1 \\ 2 \\ 3 \end{bmatrix}$ and $\mathbf{r} = \begin{bmatrix} 4 \\ 6 \\ 8 \end{bmatrix} + \mu \begin{bmatrix} 3 \\ 4 \\ 5 \end{bmatrix}$

6 Show that the following pairs of lines intersect and find the coordinates of the point of intersection.

    a $\mathbf{r} = \begin{bmatrix} 2 \\ 1 \\ -1 \end{bmatrix} + s \begin{bmatrix} 1 \\ -1 \\ 2 \end{bmatrix}$ and $\mathbf{r} = \begin{bmatrix} 1 \\ 2 \\ -5 \end{bmatrix} + t \begin{bmatrix} -1 \\ 1 \\ -1 \end{bmatrix}$

    b $\mathbf{r} = \begin{bmatrix} 4 \\ -1 \\ 2 \end{bmatrix} + \lambda \begin{bmatrix} 2 \\ 2 \\ -5 \end{bmatrix}$ and $\mathbf{r} = \begin{bmatrix} 3 \\ -5 \\ 6 \end{bmatrix} + \mu \begin{bmatrix} 1 \\ -2 \\ -1 \end{bmatrix}$

7 a Show that the points $A(-3, 6, 0)$ and $B(7, -9, -5)$ lie on the line with equation

$$\mathbf{r} = \begin{bmatrix} 1 \\ 0 \\ -2 \end{bmatrix} + \lambda \begin{bmatrix} 2 \\ -3 \\ -1 \end{bmatrix}.$$

  b Find the length of AB.

8 a Find a vector equation of the line l, which passes through the points $A(2, 1, -2)$ and $B(-3, 4, 1)$.

  b Given that the point $C(p, -11, q)$ lies on l, find p and q.

SKILLS CHECK   **7C EXTRA is on the CD**

7.4 Scalar product

The scalar product. Its use for calculating the angle between two lines.

The **scalar product** of two vectors **a** and **b** is **a.b** and is calculated as follows:

If $\mathbf{a} = \begin{bmatrix} a_1 \\ a_2 \\ a_3 \end{bmatrix}$ and $\mathbf{b} = \begin{bmatrix} b_1 \\ b_2 \\ b_3 \end{bmatrix}$ then

$$\mathbf{a.b} = \begin{bmatrix} a_1 \\ a_2 \\ a_3 \end{bmatrix} . \begin{bmatrix} b_1 \\ b_2 \\ b_3 \end{bmatrix} = a_1 b_1 + a_2 b_2 + a_3 b_3$$

Note:
a.b is said 'a dot b'. It is sometimes called the dot product.

For example, if $\mathbf{a} = \begin{bmatrix} 2 \\ -3 \\ 4 \end{bmatrix}$ and $\mathbf{b} = \begin{bmatrix} 5 \\ 3 \\ 2 \end{bmatrix}$, then

$$\mathbf{a}.\mathbf{b} = \begin{bmatrix} 2 \\ -3 \\ 4 \end{bmatrix} . \begin{bmatrix} 5 \\ 3 \\ 2 \end{bmatrix} = 2 \times 5 + (-3) \times 3 + 4 \times 2 = 9$$

Note:
It is called the scalar product because it gives a scalar result.

Finding the angle between two lines

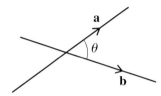

If θ is the angle between the vectors when the vectors converge to a point or diverge from a point, then θ can be found using the formula

$$\mathbf{a}.\mathbf{b} = |\mathbf{a}||\mathbf{b}|\cos\theta$$

If $\mathbf{a}$ and $\mathbf{b}$ are parallel then

$$\mathbf{a}.\mathbf{b} = |\mathbf{a}||\mathbf{b}|\cos 0° = |\mathbf{a}||\mathbf{b}|$$

If $\mathbf{a}$ and $\mathbf{b}$ are perpendicular then

$$\mathbf{a}.\mathbf{b} = |\mathbf{a}||\mathbf{b}|\cos 90° = 0$$

In reverse, if $\mathbf{a}.\mathbf{b} = 0$, then two non-zero vectors $\mathbf{a}$ and $\mathbf{b}$ are perpendicular.

Note:
$\mathbf{a}.\mathbf{b} = |\mathbf{a}||\mathbf{b}|\cos\theta$
$= |\mathbf{b}||\mathbf{a}|\cos\theta = \mathbf{b}.\mathbf{a}$

Note:
If $\mathbf{a}$ and $\mathbf{b}$ are position vectors then $\theta = \angle AOB$.

Recall:
$\cos 0° = 1$, $\cos 90° = 0$
(C2 Section 4.5).

Example 7.10 Find the angle between the vectors $\mathbf{a}$ and $\mathbf{b}$ when

a $\mathbf{a} = \begin{bmatrix} 8 \\ -1 \\ 4 \end{bmatrix}, \mathbf{b} = \begin{bmatrix} 2 \\ 1 \\ -1 \end{bmatrix}$ **b** $\mathbf{a} = \begin{bmatrix} 4 \\ 5 \\ 3 \end{bmatrix}, \mathbf{b} = \begin{bmatrix} 3 \\ -5 \\ -4 \end{bmatrix}$

Step 1: Calculate the scalar product $\mathbf{a}.\mathbf{b}$.

a $\mathbf{a}.\mathbf{b} = \begin{bmatrix} 8 \\ -1 \\ 4 \end{bmatrix} . \begin{bmatrix} 2 \\ 1 \\ -1 \end{bmatrix} = 8 \times 2 + (-1) \times 1 + 4 \times (-1) = 11$

Tip:
Show some working in case you make a slip.

Step 2: Find the modulus of each vector.

$|\mathbf{a}| = \sqrt{8^2 + (-1)^2 + 4^2} = \sqrt{81}$

$|\mathbf{b}| = \sqrt{2^2 + 1^2 + (-1)^2} = \sqrt{6}$

Recall:
Modulus of a vector (Section 7.1).

Step 3: Use the scalar product formula connecting $\mathbf{a}$, $\mathbf{b}$ and θ.

$\mathbf{a}.\mathbf{b} = |\mathbf{a}||\mathbf{b}|\cos\theta$

$11 = \sqrt{81}\sqrt{6}\cos\theta$

Step 4: Rearrange to find θ.

$\cos\theta = \dfrac{11}{\sqrt{81}\sqrt{6}}$

$\theta = 60.1°$ (1 d.p.)

Tip:
Leave your answers in surd form: using decimals at this stage could lead to an accuracy later.

Tip:
If the question doesn't state the required degree of accuracy, give angles to 1 d.p.

Step 1: Calculate the scalar product **a.b**.

b $\mathbf{a.b} = \begin{bmatrix} 4 \\ 5 \\ 3 \end{bmatrix} \cdot \begin{bmatrix} 3 \\ -5 \\ -4 \end{bmatrix} = 4 \times 3 + 5 \times (-5) + 3 \times (-4) = -25$

Tip:
The scalar product gives a scalar result so your answer should be a number.

Step 2: Find the modulus of each vector.

$|\mathbf{a}| = \sqrt{4^2 + 5^2 + 3^2} = \sqrt{50}$
$|\mathbf{b}| = \sqrt{3^2 + (-5)^2 + (-4)^2} = \sqrt{50}$

Note:
In part **a**, $\cos \theta$ is positive, so the angle between the vectors was acute. In part **b**, $\cos \theta$ is negative, so the angle is obtuse.

Step 3: Use the scalar product formula connecting **a**, **b** and θ.

$\mathbf{a.b} = |\mathbf{a}||\mathbf{b}| \cos \theta$
$-25 = \sqrt{50} \sqrt{50} \cos \theta$

Step 4: Rearrange to find θ.

$\cos \theta = -\dfrac{25}{\sqrt{50}\sqrt{50}}$

$\theta = 120°$

Example 7.11 The vector $\mathbf{a} = \begin{bmatrix} 6 \\ a \\ 5 \end{bmatrix}$ is perpendicular to the vector $\mathbf{b} = \begin{bmatrix} 2 \\ 1 \\ -2 \end{bmatrix}$. Find a.

Step 1: Calculate **a.b**.

$\mathbf{a.b} = \begin{bmatrix} 6 \\ a \\ 5 \end{bmatrix} \cdot \begin{bmatrix} 2 \\ 1 \\ -2 \end{bmatrix} = 6 \times 2 + a \times 1 + 5 \times (-2) = 12 + a - 10 = a + 2$

Step 2: Use the scalar product property relating to perpendicular vectors.

If **a** is perpendicular to **b**, then $\mathbf{a.b} = 0$
$\Rightarrow \quad a + 2 = 0$

Recall:
Perpendicular vectors have a zero scalar product.

Step 3: Solve for a.

$a = -2$

Example 7.12 The line l_1 has equation $\begin{bmatrix} x \\ y \\ z \end{bmatrix} = \begin{bmatrix} 1 \\ 2 \\ -1 \end{bmatrix} + \lambda \begin{bmatrix} 3 \\ -1 \\ 4 \end{bmatrix}$.

The line l_2 has equation $\begin{bmatrix} x \\ y \\ z \end{bmatrix} = \begin{bmatrix} 2 \\ 3 \\ -1 \end{bmatrix} + \mu \begin{bmatrix} -2 \\ 3 \\ 1 \end{bmatrix}$.

Recall:
$\mathbf{r} = \begin{bmatrix} x \\ y \\ z \end{bmatrix}$

Find the acute angle between the lines l_1 and l_2, giving your answer to the nearest $0.1°$.

Step 1: Calculate the scalar product of the direction vectors of the lines.

The lines have direction vectors $\mathbf{d_1} = \begin{bmatrix} 3 \\ -1 \\ 4 \end{bmatrix}$ and $\mathbf{d_2} = \begin{bmatrix} -2 \\ 3 \\ 1 \end{bmatrix}$.

$\mathbf{d_1.d_2} = \begin{bmatrix} 3 \\ -1 \\ 4 \end{bmatrix} \cdot \begin{bmatrix} -2 \\ 3 \\ 1 \end{bmatrix} = 3 \times (-2) + (-1) \times 3 + 4 \times 1 = -5$

Tip:
Always calculate $\mathbf{d_1.d_2}$ first. If this is zero the lines are perpendicular and you need go no further.

Step 2: Find the modulus of each direction vector.

$|\mathbf{d_1}| = \sqrt{3^2 + (-1)^2 + 4^2} = \sqrt{26}$
$|\mathbf{d_2}| = \sqrt{(-2)^2 + 3^2 + 1^2} = \sqrt{14}$

Step 3: Use the scalar product formula.

$\cos \theta = \dfrac{\mathbf{d_1.d_2}}{|\mathbf{d_1}||\mathbf{d_2}|} = \dfrac{-5}{\sqrt{26}\sqrt{14}} = -0.2620\ldots$

Note:
Since $\cos \theta$ is negative, you have calculated the obtuse angle between the lines.

Step 4: Find θ.

$\theta = 105.19\ldots°$

Step 5: Calculate the acute angle if necessary.

Required angle $= 180° - 105.19\ldots° = 74.81\ldots°$

The acute angle between l_1 and l_2 is $74.8°$ (1 d.p.).

Note:
Giving the angle to the nearest $0.1°$ is the same as giving it in degrees to one decimal place.

Shortest distance from a point to a line

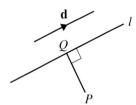

Consider the line l with vector equation $\mathbf{r} = \mathbf{a} + t\mathbf{d}$.

The shortest distance from P to the line l is the distance PQ, where Q lies on the line and PQ is perpendicular to l.

Note: Q is the **foot of the perpendicular** from P to the line.

To find the coordinates of Q, and hence the distance PQ, first let Q be the point where $t = q$.

Find $\overrightarrow{OQ}$ and $\overrightarrow{PQ}$ in terms of q. Then find $\overrightarrow{PQ}.\mathbf{d}$, also in terms of q.

Recall: $\mathbf{d}$ is the direction vector of l.

Since PQ is perpendicular to the line l, set $\overrightarrow{PQ}.\mathbf{d} = 0$. This will give you the value of q. Substitute this into $\overrightarrow{OQ}$ to find the position vector of Q and hence the coordinates of Q.

Tip: You can just substitute into $\overrightarrow{OQ}$.

The distance PQ can now be calculated.

This procedure is illustrated in the following example.

Example 7.13 The line l has equation $\begin{bmatrix} x \\ y \\ z \end{bmatrix} = \begin{bmatrix} 1 \\ 1 \\ -3 \end{bmatrix} + t \begin{bmatrix} 3 \\ 1 \\ 4 \end{bmatrix}$.

The point P has coordinates $(5, -7, 9)$.

a The point Q on the line l is where $t = q$. Show that
$$\overrightarrow{PQ}.\begin{bmatrix} 3 \\ 1 \\ 4 \end{bmatrix} = 26q - 52$$

b Hence find the coordinates of the foot of the perpendicular from the point P to the line l.

c Hence find the shortest distance from P to the line l, leaving your answer in simplified surd form.

Step 1: State the position vectors of P and Q.

a $\overrightarrow{OP} = \mathbf{p} = \begin{bmatrix} 5 \\ -7 \\ 9 \end{bmatrix}$

$\overrightarrow{OQ} = \mathbf{q} = \begin{bmatrix} 1 \\ 1 \\ -3 \end{bmatrix} + q \begin{bmatrix} 3 \\ 1 \\ 4 \end{bmatrix} = \begin{bmatrix} 1 + 3q \\ 1 + q \\ -3 + 4q \end{bmatrix}$

Note: Both the position vector of Q and the vector $\overrightarrow{PQ}$ will be in terms of q at this stage.

Step 2: Find $\overrightarrow{PQ}$ in terms of q.

$\overrightarrow{PQ} = \mathbf{q} - \mathbf{p} = \begin{bmatrix} 1 + 3q \\ 1 + q \\ -3 + 4q \end{bmatrix} - \begin{bmatrix} 5 \\ -7 \\ 9 \end{bmatrix} = \begin{bmatrix} 1 + 3q - 5 \\ 1 + q + 7 \\ -3 + 4q - 9 \end{bmatrix} = \begin{bmatrix} 3q - 4 \\ q + 8 \\ 4q - 12 \end{bmatrix}$

Step 3: Find the required scalar product.

$\overrightarrow{PQ}.\begin{bmatrix} 3 \\ 1 \\ 4 \end{bmatrix} = \begin{bmatrix} 3q - 4 \\ q + 8 \\ 4q - 12 \end{bmatrix}.\begin{bmatrix} 3 \\ 1 \\ 4 \end{bmatrix}$

Note: $\begin{bmatrix} 3 \\ 1 \\ 4 \end{bmatrix}$ is the direction vector of the line l.

$= 3(3q - 4) + q + 8 + 4(4q - 12)$
$= 9q - 12 + q + 8 + 16q - 48$
$= 26q - 52$

Step 1: Apply the scalar product rule for perpendicular lines.

b When $\overrightarrow{PQ}$ is perpendicular to l,

$$\overrightarrow{PQ} \cdot \begin{bmatrix} 3 \\ 1 \\ 4 \end{bmatrix} = 0 \Rightarrow 26q - 52 = 0 \Rightarrow q = 2$$

Step 2: Substitute the value for q into $\overrightarrow{OQ}$.

When $q = 2$, $\overrightarrow{OQ} = \begin{bmatrix} 1 + 3q \\ 1 + q \\ -3 + 4q \end{bmatrix} = \begin{bmatrix} 1 + 6 \\ 1 + 2 \\ -3 + 8 \end{bmatrix} = \begin{bmatrix} 7 \\ 3 \\ 5 \end{bmatrix}$

Note:
This fixes a particular position for Q at the foot of the perpendicular from P to the line.

Step 3: State the coordinates of Q.

The foot of the perpendicular from P to the line l has coordinates (7, 3, 5).

Step 1: Substitute $q = 2$ into $\overrightarrow{PQ}$.

c $\overrightarrow{PQ} = \begin{bmatrix} 3q - 4 \\ q + 8 \\ 4q - 12 \end{bmatrix} = \begin{bmatrix} 3 \times 2 - 4 \\ 2 + 8 \\ 4 \times 2 - 12 \end{bmatrix} = \begin{bmatrix} 2 \\ 10 \\ -4 \end{bmatrix}$

Tip:
Alternatively use
$d = \sqrt{(x_2 - x_1)^2 + (y_2 - y_1)^2 + (z_2 - z_1)^2}$
with $P(5, -7, 9)$ and $Q(7, 3, 5)$.

Step 2: Calculate $|\overrightarrow{PQ}|$.

$|\overrightarrow{PQ}| = \sqrt{2^2 + 10^2 + (-4)^2} = \sqrt{120} = 2\sqrt{30}$

The shortest distance from P to l is $2\sqrt{30}$.

SKILLS CHECK 7D: Scalar product

1 Find the angle between the following pairs of vectors.

a $\mathbf{a} = \begin{bmatrix} 2 \\ 0 \\ 1 \end{bmatrix}$, $\mathbf{b} = \begin{bmatrix} -3 \\ -5 \\ 2 \end{bmatrix}$ **b** $\mathbf{a} = \begin{bmatrix} -2 \\ 1 \\ -1 \end{bmatrix}$, $\mathbf{b} = \begin{bmatrix} -2 \\ -1 \\ 3 \end{bmatrix}$ **c** $\mathbf{a} = \begin{bmatrix} 3 \\ 2 \\ -4 \end{bmatrix}$, $\mathbf{b} = \begin{bmatrix} 1 \\ -3 \\ -2 \end{bmatrix}$

2 Given that $\overrightarrow{OA} = \begin{bmatrix} 3 \\ -7 \\ 1 \end{bmatrix}$ and $\overrightarrow{OB} = \begin{bmatrix} 4 \\ 1 \\ -5 \end{bmatrix}$, show that $\overrightarrow{OA}$ and $\overrightarrow{OB}$ are perpendicular.

3 Use a vector method to find the area of triangle AOB given that O is the origin, A is the point $(1, -3, 2)$ and B is the point $(-4, -1, 3)$. Give your answer to three significant figures.

4 Find the acute angle between each of the following pairs of lines.

a $\mathbf{r} = \begin{bmatrix} 1 \\ 2 \\ -2 \end{bmatrix} + s \begin{bmatrix} 2 \\ -3 \\ 6 \end{bmatrix}$ and $\mathbf{r} = \begin{bmatrix} 6 \\ 4 \\ -3 \end{bmatrix} + t \begin{bmatrix} 1 \\ 2 \\ -2 \end{bmatrix}$

b $\begin{bmatrix} x \\ y \\ z \end{bmatrix} = \begin{bmatrix} 7 \\ -8 \\ -2 \end{bmatrix} + \lambda \begin{bmatrix} 3 \\ 2 \\ 6 \end{bmatrix}$ and $\begin{bmatrix} x \\ y \\ z \end{bmatrix} = \begin{bmatrix} 3 \\ -1 \\ 4 \end{bmatrix} + \mu \begin{bmatrix} 1 \\ 2 \\ 3 \end{bmatrix}$

5 Given that $\mathbf{a} = \begin{bmatrix} 0 \\ p \\ 1 \end{bmatrix}$, $\mathbf{b} = \begin{bmatrix} 3 \\ 2 \\ 0 \end{bmatrix}$ and that the angle between the vectors is $60°$, find the exact value of p.

6 Use a vector method to find all the angles of triangle ABC, where A, B and C are the points $(4, -1, 4)$, $(-2, 3, 5)$ and $(1, 0, -6)$, respectively. Give your answers to one decimal place.

7 Given that $\overrightarrow{OP} = \begin{bmatrix} 2 \\ 3 \\ 5 \end{bmatrix}$, $\overrightarrow{OQ} = \begin{bmatrix} 3 \\ -1 \\ 2 \end{bmatrix}$ and $\overrightarrow{OR} = \begin{bmatrix} 10 \\ -3 \\ 7 \end{bmatrix}$,

a show that $\overrightarrow{QP}$ is perpendicular to $\overrightarrow{QR}$,

b find the area of triangle PQR, leaving your answer in simplified surd form.

8 The lines with equations $\mathbf{r} = \begin{bmatrix} 2 \\ 1 \\ 0 \end{bmatrix} + \lambda \begin{bmatrix} -1 \\ 4 \\ 2 \end{bmatrix}$ and $\mathbf{r} = \begin{bmatrix} 1 \\ 5 \\ 2 \end{bmatrix} + \mu \begin{bmatrix} 2 \\ 3 \\ -5 \end{bmatrix}$ intersect at the point X.

  a Show that the lines are perpendicular.
  b Find the coordinates of the point X.

9 The vectors $\mathbf{a} = \begin{bmatrix} a \\ -5 \\ -3 \end{bmatrix}$ and $\mathbf{b} = \begin{bmatrix} 2a \\ a \\ -1 \end{bmatrix}$ are perpendicular. Find the possible values of a.

10 The point P is on the line l, with equation $\mathbf{r} = \begin{bmatrix} 1 \\ -3 \\ 0 \end{bmatrix} + \lambda \begin{bmatrix} 2 \\ 1 \\ -2 \end{bmatrix}$. Given that $\overrightarrow{OP}$, the position vector of P, is perpendicular to the line l, find the coordinates of P.

11 P is the point $(-1, 3, 4)$.

   The line l has equation $\mathbf{r} = \begin{bmatrix} 2 \\ 1 \\ 4 \end{bmatrix} + \lambda \begin{bmatrix} 3 \\ 1 \\ 5 \end{bmatrix}$ and A is the point on the line where $\lambda = 1$.

   B is a point on the line where $\lambda = b$ and angle PBA is 90°.

   a Find the coordinates of A and hence find the distance PA.
   b i Find the value of b.
     ii Hence find the shortest distance from P to the line, giving your answer to three significant figures.

12 The line l passes through $A(-1, -5, 9)$ and $B(11, 13, -9)$.

   a Show that a vector equation of the line l is

   $$\mathbf{r} = \begin{bmatrix} -1 \\ -5 \\ 9 \end{bmatrix} + t \begin{bmatrix} 2 \\ 3 \\ -3 \end{bmatrix}$$

   b The point P, where $t = p$, lies on the line l.

   Show that $\overrightarrow{OP} \cdot \begin{bmatrix} 2 \\ 3 \\ -3 \end{bmatrix} = 22p - 44$.

   c Given that OP is perpendicular to the line l, find the coordinates of P.
   d Find the area of triangle OAB.

SKILLS CHECK **7D EXTRA** is on the CD

Examination practice 7: Vectors

1 The line l_1 has equation $\mathbf{r} = \begin{bmatrix} 7 \\ p \\ -9 \end{bmatrix} + \lambda \begin{bmatrix} 3 \\ -5 \\ -4 \end{bmatrix}$ and the line l_2 has equation $\mathbf{r} = \begin{bmatrix} 0 \\ 2 \\ -8 \end{bmatrix} + \mu \begin{bmatrix} 4 \\ 5 \\ 3 \end{bmatrix}$,

where λ and μ are parameters and p is a constant. The two lines intersect at the point A.
   a Find the value of p.
   b Find the position vector of A.
   c Prove that the angle between l_1 and l_2 is 120°.

2 The line l_1 has equation $\mathbf{r} = \begin{bmatrix} 1 \\ 0 \\ -2 \end{bmatrix} + \lambda \begin{bmatrix} 1 \\ 4 \\ 3 \end{bmatrix}$.

The line l_2 has equation $\mathbf{r} = \begin{bmatrix} 5 \\ 5 \\ 10 \end{bmatrix} + \mu \begin{bmatrix} 2 \\ -3 \\ 6 \end{bmatrix}$.

   a Show that the lines l_1 and l_2 intersect at a point P and find the position vector of P.

   b Find the acute angle between the lines l_1 and l_2, giving your answer to the nearest degree.

   c The line l_3 passes through the points $(0, 0, 0)$ and $(2, 8, 6)$.
Show that l_1 and l_3 are parallel lines.   [AQA June 2003]

3 **a** Find the vector equation of the line l_1, which passes through the points $A(3, -1, 2)$ and $B(2, 0, 2)$.

   b The line l_2 has vector equation $\mathbf{r} = \begin{bmatrix} 4 \\ 1 \\ -1 \end{bmatrix} + \mu \begin{bmatrix} 1 \\ 0 \\ -1 \end{bmatrix}$.

Show that the lines l_1 and l_2 intersect and find the coordinates of their point of intersection.

   c Show that the point $C(9, 1, -6)$ lies on the line l_2.

   d Find the coordinates of the point D on l_1 such that CD is perpendicular to l_1.   [AQA June 2004]

4 The line l_1 has vector equation

$$\mathbf{r} = \begin{bmatrix} 3 \\ 1 \\ 2 \end{bmatrix} + \lambda \begin{bmatrix} 1 \\ -1 \\ 4 \end{bmatrix}$$

and the line l_2 has vector equation

$$\mathbf{r} = \begin{bmatrix} 0 \\ 4 \\ -2 \end{bmatrix} + \mu \begin{bmatrix} 1 \\ -1 \\ 0 \end{bmatrix},$$

where λ and μ are parameters.

The lines l_1 and l_2 intersect at the point B and the acute angle between l_1 and l_2 is θ.

   a Find the coordinates of B.

   b Find the value of θ, giving your answer to the nearest degree.

5 The line l_1 has equation $\begin{bmatrix} x \\ y \\ z \end{bmatrix} = \begin{bmatrix} 3 \\ -2 \\ 1 \end{bmatrix} + t \begin{bmatrix} 4 \\ 4 \\ 3 \end{bmatrix}$.

The line l_2 has equation $\begin{bmatrix} x \\ y \\ z \end{bmatrix} = \begin{bmatrix} 8 \\ -1 \\ 2 \end{bmatrix} + s \begin{bmatrix} -1 \\ 3 \\ 2 \end{bmatrix}$.

   a The vector $\begin{bmatrix} a \\ b \\ -16 \end{bmatrix}$ is perpendicular to both l_1 and l_2. Find the values of a and b.

   b Show that the lines l_1 and l_2 intersect and find the coordinates of their point of intersection.

   c Find the acute angle between l_1 and l_2.

6 The line l_1 passes through the point $A(2, 0, 2)$ and has equation

$$\mathbf{r} = \begin{bmatrix} 2 \\ 0 \\ 2 \end{bmatrix} + t \begin{bmatrix} 2 \\ 6 \\ -3 \end{bmatrix}.$$

a Show that the line l_2 which passes through the points $B(4, 4, -5)$ and $C(0, -8, 1)$ is parallel to the line l_1.

b P is the point on line l_1 such that angle CPA is a right angle.
   i Find the value of the parameter t at P.
   ii Hence show that the shortest distance between the lines l_1 and l_2 is $2\sqrt{5}$.   [AQA June 2002]

  7 The line l_1 has equation $\mathbf{r} = \begin{bmatrix} 3 \\ 1 \\ -2 \end{bmatrix} + s \begin{bmatrix} 1 \\ -1 \\ 4 \end{bmatrix}$.

The line l_2 has equation $\mathbf{r} = \begin{bmatrix} 1 \\ 0 \\ 5 \end{bmatrix} + t \begin{bmatrix} -2 \\ -3 \\ 1 \end{bmatrix}$.

a Show that the two lines are skew.

b Find the acute angle between the directions of the two lines.

8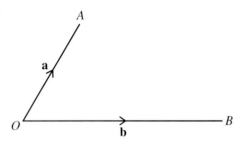

The diagram shows points O, A and B, with $\overrightarrow{OA} = \mathbf{a}$ and $\overrightarrow{OB} = \mathbf{b}$.

a Make a sketch of the diagram and mark the points C and D such that $\overrightarrow{OC} = \mathbf{a} + 2\mathbf{b}$ and $\overrightarrow{OD} = 2\mathbf{a} + \mathbf{b}$.

b Find $\overrightarrow{AB}$, $\overrightarrow{BC}$, $\overrightarrow{DC}$ and $\overrightarrow{AD}$, simplifying your answers where possible.

c Prove that $ABCD$ is a parallelogram.

9 The point A has coordinates $(-3, 0, -7)$ and the point B has coordinates $(-1, 4, 1)$.

The line l has equation $\mathbf{r} = \begin{bmatrix} 1 \\ 2 \\ -1 \end{bmatrix} + \lambda \begin{bmatrix} 2 \\ 1 \\ 3 \end{bmatrix}$.

a Show that the point A lies on the line l.

b Find the acute angle between the line l and $\overrightarrow{AB}$, giving your answer to three significant figures.

10 The line l_1 passes through $A(2, 1, -3)$ and $B(3, 0, -2)$.

The line l_2 has equation $\mathbf{r} = \begin{bmatrix} 6 \\ 4 \\ -5 \end{bmatrix} + t \begin{bmatrix} 4 \\ 3 \\ -2 \end{bmatrix}$.

    a Find a vector equation of the line l_1.

    b Show that A lies on l_2.

    c Find the acute angle between the lines l_1 and l_2, giving your answer to the nearest degree.

 11 The line l passes through $A(0, 4, -2)$ and $B(3, -2, 7)$. The point P lies on the line l and OP is perpendicular to l where O is the origin.

    a Find a vector equation of the line l.

    b Find the coordinates of P.

    c Find the area of the triangle OAP.

Practice exam paper

Answer **all** questions.
Time allowed: 1 hour 30 minutes
You may use a graphics calculator

1 a Express $10 \cos x - 2 \sin x$ in the form $R \cos(x + \alpha)$, where $R > 0$ and $0° < \alpha < 90°$.
Give the value of α to the nearest $0.1°$. *(3 marks)*

b Solve the equation $10 \cos x - 2 \sin x = 5$ for $0° \leq x \leq 360°$ giving your answers to one decimal place. *(4 marks)*

c Write down the minimum value of $10 \cos x - 2 \sin x$. *(1 mark)*

2 A curve is defined by the parametric equations
$$x = 3 + 2t, \quad y = 1 + \frac{4}{t}$$

a Find $\frac{dy}{dx}$ in terms of t. *(4 marks)*

b Find an equation of the normal to the curve at the point where $t = 4$. *(4 marks)*

c Show that the Cartesian equation of the curve can be written in the form
$$(x - 3)(y - 1) = k$$ *(3 marks)*

3 The polynomial f(x) is defined by $f(x) = 6x^3 + 7x^2 - x - 2$.

a Use the remainder theorem to find the remainder when f(x) is divided by $3x - 2$. *(2 marks)*

b Find $f(-1)$. *(1 mark)*

c Simplify fully $\dfrac{(x + 1)(3x + 2)}{6x^3 + 7x^2 - x - 2}$. *(4 marks)*

4 a Express $\dfrac{10x - 1}{(x + 2)(2x - 3)}$ in the form $\dfrac{A}{x + 2} + \dfrac{B}{2x - 3}$. *(3 marks)*

b Hence find $\displaystyle\int \dfrac{10x - 1}{(x + 2)(2x - 3)} \, dx$. *(3 marks)*

5 a Obtain the binomial expansion of $(1 + x)^{\frac{1}{4}}$ up to and including the term in x^2. *(2 marks)*

b Show that
$$(16 + 5x)^{\frac{1}{4}} \approx 2 + \frac{5}{32}x - \frac{75}{4096}x^2$$
for small values of x. *(3 marks)*

c Use your answer to part **b** with $x = -\frac{1}{5}$ to show that $\sqrt[4]{15} \approx \dfrac{8061}{4096}$. *(3 marks)*

6 a Express $\cos 2x$ in terms of $\sin x$ and $\cos x$. *(1 mark)*

b Hence show that $\dfrac{\cos 2x}{\cos x + \sin x} = \cos x - \sin x$. *(2 marks)*

c Hence find $\displaystyle\int_0^{\frac{\pi}{4}} \dfrac{\cos 2x \cos x}{\cos x + \sin x} \, dx$. *(6 marks)*

7 a The population, P, of insects in a colony is given by $P = Ae^{kt}$, where A and k are constants and the time t is measured in months.
Given that $P = 600$ when $t = 0$ and that $P = 800$ when $t = 10$, find the rate of increase of P when $P = 1000$. *(5 marks)*

b Solve the differential equation
$$\frac{dy}{dx} = (2 - x)y^3$$
given that $y = 1$ when $x = 2$. Give your answer in the form $y^2 = f(x)$. *(6 marks)*

8 The line l_1 has vector equation $\mathbf{r} = \begin{bmatrix} 4 \\ 3 \\ -5 \end{bmatrix} + \lambda \begin{bmatrix} 1 \\ 1 \\ -4 \end{bmatrix}$.

The line l_2 has vector equation $\mathbf{r} = \begin{bmatrix} -2 \\ -5 \\ 5 \end{bmatrix} + \mu \begin{bmatrix} 2 \\ 3 \\ -1 \end{bmatrix}$.

a Show that the lines l_1 and l_2 intersect at a point A and find the position vector of A. *(4 marks)*

b The point B has coordinates $(6, 7, 1)$. Verify that B lies on l_2. *(2 marks)*

c The point P on the line l_1 is where $\lambda = p$.

i Show that $\overrightarrow{BP} \cdot \begin{bmatrix} 1 \\ 1 \\ -4 \end{bmatrix} = 18p + 18$. *(4 marks)*

ii C is the foot of the perpendicular from the point B to the line l_1.
Find the distance between the points A and C. *(5 marks)*

Answers

SKILLS CHECK 1A (page 6)

1. **a** $\dfrac{x}{3x-2}$ **b** $\dfrac{x+2}{x+5}$ **c** $\dfrac{2}{x-3}$ **d** $\dfrac{3x+1}{2x-3}$
2. **a** $x(x-7)$ **b** $\dfrac{x+4}{x}$ **c** $\dfrac{x-1}{2}$ **d** $\dfrac{t}{3}$
3. **a** $\dfrac{8x+7}{(x-1)(x+2)}$ **b** $\dfrac{14-3y}{4-y}$
   c $\dfrac{x-8}{(x-2)(x-5)}$ **d** $\dfrac{2x-3}{(x-3)(x-1)}$
4. **a** $x^2 + x + 1$ **b** $x = -2$ or 1
5. $3 + \dfrac{7}{x-1}$
6. $x - 1 - \dfrac{2}{3x-4}$
7. $A = 2, B = 1, C = 2$
8. **a** $a = 2$ **b** $-\dfrac{29}{9}$
9. **a** 0 **b** $(3x-1)(2x-5)(x+2)$
   c $x = \tfrac{1}{3}$ or $\tfrac{5}{2}$ or -2

SKILLS CHECK 1B (page 11)

1. **a** $\dfrac{1}{x-3} - \dfrac{1}{x+1}$ **b** $\dfrac{1}{2(x-3)} - \dfrac{1}{2(3x-5)}$
   c $\dfrac{3}{x+2} - \dfrac{2}{2x-3}$
2. **a** $\dfrac{1}{x-1} - \dfrac{1}{x} - \dfrac{1}{x^2}$ **b** $\dfrac{11}{x+2} - \dfrac{9}{x+1} + \dfrac{5}{(x+1)^2}$
   c $\dfrac{2}{2x-1} + \dfrac{1}{x+2} - \dfrac{3}{(x+2)^2}$
3. **a** $\dfrac{5}{x-3} - \dfrac{4}{x+3}$ **b** $\dfrac{3}{4(1-x)} + \dfrac{3}{4(1+x)} - \dfrac{3}{2(1+x)^2}$
4. **a** $\dfrac{2}{1-2x} + \dfrac{1}{x+1}$ **b** $(-\tfrac{1}{4}, \tfrac{8}{3})$
5. **a** $\dfrac{2}{1-x} - \dfrac{3}{1+x}$ **b** **i** $\dfrac{2}{(1-x)^2} + \dfrac{3}{(1+x)^2}$ **ii** -2
6. **a** $-\dfrac{1}{x-5} - \dfrac{2}{2x+3}$
7. $5 - \dfrac{3}{2+x} + \dfrac{2}{1-x}$
8. $1 - \dfrac{4}{x-4} + \dfrac{12}{x-6}$

Exam practice 1 (page 12)

1. $\dfrac{3x}{2x+1}$
2. **a** $\dfrac{21-x}{(x+3)(x-8)}$ **b** $9, -5$
3. **a** $x(x-1)(x+1)$ **b** $\dfrac{2}{x} - \dfrac{3}{x+1} + \dfrac{1}{x-1}$
4. **a** $\dfrac{2}{2x-1} - \dfrac{1}{x+3}$
   b **i** $\dfrac{1}{(x+3)^2} - \dfrac{4}{(2x-1)^2}$ **ii** $\dfrac{16}{(2x-1)^3} - \dfrac{2}{(x+3)^3}$
5. $\dfrac{2}{x+2} - \dfrac{1}{(x-1)^2}$
6. $3 + \dfrac{2}{x-1} + \dfrac{5}{2x+1}$
7. $-2 + x + \dfrac{4}{x+1} + \dfrac{2}{x+4}$; $A = -2, B = 1, C = 4, D = 2$

8. **a** **i** 12.5 **ii** 0
   b $(2x-1)(x-2)(x+3)$ **c** $x = \tfrac{1}{2}$ or 2 or -3
9. **b** $(3x-1)(x^2+2x+2)$ **c** $x = \tfrac{1}{3}$

SKILLS CHECK 2A (page 16)

1. **a** $(0, 2)$ **b** $(0, -4), (0, 4)$
2. **a** $(6, 0)$ **b** $(-1, 0), (5, 0)$
3. $a = 4$
4. **a** $y^2 = 4x$ **b** $\dfrac{x^2}{a^2} + \dfrac{y^2}{b^2} = 1$
   c $y = \dfrac{1}{x-3}$ **d** $y^2 = \dfrac{1}{x^2} - 1$
5. $(\tfrac{3}{5}, \tfrac{2}{5}), (\tfrac{3}{2}, -\tfrac{1}{2})$
6. $y^2 = 9(x^2 - 1)$
7. **a** $(x-2)^2 + (y+4)^2 = 9$ **b** Radius 3, centre $(2, -4)$
8. **a** $x + y = 2t, x - y = \dfrac{2}{t}$ **b** $(x+y)(x-y) = 4$

Exam practice 2 (page 16)

1. $c = 3$
2. **a** $y = (x+1)^3 + (x+1)^2$ **b** $(-1, 0), (-2, 0)$
3. $k = 3$
4. **a** $(x+1)^2 + (y-5)^2 = 16$ **b** Radius 4, centre $(-1, 5)$
5. $y + 2x = 7$

SKILLS CHECK 3A (page 22)

1. **a** $1 + 2x + 3x^2 + 4x^3, |x| < 1$ **b** $1 - 3x - 9x^2 - 45x^3, |x| < \tfrac{1}{9}$
   c $1 + 6x + 30x^2 + 140x^3, |x| < \tfrac{1}{4}$
2. **a** $-1 - 6x - 24x^2, |x| < \tfrac{1}{2}$ **b** $\tfrac{1}{8} - \tfrac{3}{64}x + \tfrac{15}{1024}x^2, |x| < 4$
3. $2 - 3x + 4x^2 - 5x^3$
4. $\tfrac{15}{8}x^2$
5. $1 + \tfrac{1}{2}x - \tfrac{1}{8}x^2 + \tfrac{1}{16}x^3, 1.7321$
6. **a** $\dfrac{1}{1+2x} + \dfrac{3}{1-x}$ **b** $4 + x + 7x^2$ **c** $|x| < \tfrac{1}{2}$
7. **a** $\dfrac{2}{2+x} - \dfrac{1}{(1-x)^2}$ **b** $-\tfrac{5}{2}x - \tfrac{11}{4}x^2 - \tfrac{33}{8}x^3$ **c** $|x| < 1$
8. **a** $a = -2, n = -1$ **b** 8 **c** $|x| < \tfrac{1}{2}$

Exam practice 3 (page 23)

1. **a** **i** $1 - x + x^2 - x^3$ **ii** $1 + 2x - 2x^2 + 4x^3$
   b $k = 4$
2. **a** $1 - x + \tfrac{3}{4}x^2 - \tfrac{1}{2}x^3$ **b** $\tfrac{5}{4}$
3. **a** $\tfrac{1}{27} - \tfrac{2}{27}x + \tfrac{8}{81}x^2 - \tfrac{80}{729}x^3$ **b** $|x| < \tfrac{3}{2}$
4. **a** $\dfrac{1}{1-x} + \dfrac{2}{2+x}$ **b** **ii** $1 + x + x^2$
   c $2 + \tfrac{1}{2}x + \tfrac{5}{4}x^2$
5. **a** **i** $1 - x + x^2 - x^3$ **b** $\dfrac{1}{1+x} + \dfrac{5}{3+2x}$
   c $\tfrac{8}{3} - \tfrac{19}{9}x + \tfrac{47}{27}x^2 - \tfrac{121}{81}x^3$
6. **a** $A = -1, B = -1$ **b** $-\tfrac{3}{2} - \tfrac{5}{4}x - \tfrac{9}{8}x^2 - \tfrac{17}{16}x^3$
   c $|x| < 1$
7. **a** $k = 2, n = -\tfrac{1}{2}$ **b** $-\tfrac{5}{2}$ **c** $|x| < \tfrac{1}{2}$

SKILLS CHECK 4A (page 30)

1. **a** $\tfrac{3}{5}$ **b** $\tfrac{5}{13}$ **c** $\tfrac{56}{65}$ **d** $\tfrac{65}{56}$
2. **a** $\dfrac{1}{\sqrt{2}}$ **c** $45°, 225°$
3. See CD.
4. **a** $\sin(X - Y) = \sin X \cos Y - \cos X \sin Y$
5. $\tfrac{1}{2}$
6. See CD.
7. **a** $\sqrt{5}\cos(x - 26.6°)$ **b** $90°, 323.1°$ (1 d.p.)
8. **a** $f(\theta) = 5\cos(\theta + 0.64^c)$ **b** **i** 5 **ii** 5.64^c

9 **a** $\sqrt{2}\sin(x+45°)$ **b** $-75°, 165°$
10 $\alpha = 61.9°, k = 8$
11 **a** $2\sqrt{5}\cos(x - 0.464^c)$ **b** 1.30^c (2 d.p.), 5.91^c (2 d.p.)

SKILLS CHECK 4B (page 35)

1 **a** $0°, 180°, 210°, 330°, 360°$ **b** $45°, 90°, 135°, 225°, 270°, 315°$
2 **a** $0^c, 2.09^c$ (2 d.p.), 4.19^c (2 d.p.)   **b** $0^c, 4.19^c$ (2 d.p.)
3 $9.6°$ (1 d.p.), $90°, 170.4°$ (1 d.p.), $270°$
4 **a** LHS $= \dfrac{\sin A}{\cos A} + \dfrac{\cos A}{\sin A}$
   $= \dfrac{\sin^2 A + \cos^2 A}{\cos A \sin A}$
   $= \dfrac{1}{\frac{1}{2}\sin 2A}$
   $= 2\operatorname{cosec} 2A$
   $=$ RHS
   b 0.13^c (2 d.p.), 1.44^c (2 d.p.) 3.27^c (2 d.p.), 4.59^c (2 d.p.)
5 **a i** $\cos 2A = 2\cos^2 A - 1$ **ii** $\cos 2A = 1 - 2\sin^2 A$
  b See CD.
6 $0°, 60°, 120°, 180°, 240°, 300°, 360°$
7 **a** $y = 1 - 2x^2$ **b** $(0, 1), \left(-\dfrac{1}{\sqrt{2}}, 0\right), \left(\dfrac{1}{\sqrt{2}}, 0\right)$
8 $\cos 3A = 4\cos^3 A - 3\cos A$
9 **a** $\frac{1}{2}x - \frac{1}{16}\sin 8x + c$ **b** $-\frac{1}{12}\cos 6x + c$
10 $2\pi^2$
11 **a** $\cos^2 A \equiv \frac{1}{2}(1 + \cos 2A)$ **b** $\frac{1}{2}\pi$
12 **a** $x - \frac{1}{2}\cos 2x + c$ **b** $\frac{1}{2}\sin 2x + c$

Exam practice 4 (page 36)

1 **a** $-\frac{5}{13}$ **b** $\frac{33}{65}$
2 **a** LHS $= \sin\alpha\cos\beta + \cos\alpha\sin\beta + \sin\alpha\cos\beta - \cos\alpha\sin\beta$
   $= 2\sin\alpha\cos\beta$
   $=$ RHS
  b i $\sin 10x + \sin 6x$ **ii** $-\frac{3}{10}\cos 10x - \frac{1}{2}\cos 6x + c$
3 **a** $10\cos(x + 53.1°)$ **b** $19°, 234°$
4 **a** 1.176^c (3 d.p.)   **b** $26\sin(\theta + 1.176^c)$
  c i 26 **ii** 0.395^c (3 d.p.)
5 **a** $25\cos(\theta - 1.287...^c)$ **b** 0.2^c (1 d.p.), 2.3^c (1 d.p.)
6 **a i** $L = 2\sin\theta + 4\cos\theta$ **ii** $2\sqrt{5}\sin(\theta + 1.107^c)$
  b i $2\sqrt{5}$ **ii** 0.46^c (2 d.p.)
7 **a** LHS $= \dfrac{\cot^2\theta}{\operatorname{cosec}^2\theta} = \dfrac{\cos^2\theta}{\sin^2\theta} \div \dfrac{1}{\sin^2\theta}$
   $= \cos^2\theta =$ RHS
  b $14.0°$ (1 d.p.), $90°, 194.0°$ (1 d.p.), $270°$
8 $48°$ (nearest °), $120°, 240°, 312°$ (nearest °)
9 LHS $= \dfrac{1 - (1 - 2\sin^2\theta)}{1 + (2\cos^2\theta - 1)}$
   $= \dfrac{2\sin^2\theta}{2\cos^2\theta}$
   $= \tan^2\theta$
   $= \sec^2\theta - 1$
   $=$ RHS
10 $0°, \pm 43.3°$ (1 d.p.)
11 **a** $\sin^2 A = \frac{1}{2}(1 - \cos 2A)$ **b** $\frac{5}{2}x - \frac{1}{4}\sin 2x + c$
12 **a** $3\sin 2x - 2\cos 2x$
   b $16.8°$ (1 d.p.), $106.8°$ (1 d.p.), $196.8°$ (1 d.p.), $286.8°$ (1 d.p.)
13 **a** $\cos 3\theta$ **b** To nearest degree: $25°, 95°, 145°$
14 **a** $\frac{3}{2} + \frac{3}{2}\cos x$ **b** $\frac{3}{2}x + \frac{3}{2}\sin x + c$
15 **a** $2\sin A \cos A$ **b** $1 - 2\sin^2 A$
   c LHS $= \dfrac{2\sin A \cos A}{1 - (1 - 2\sin^2 A)}$
      $= \dfrac{2\sin A \cos A}{2\sin^2 A} = \cot A =$ RHS

16 **a** See CD for proof.
   b i $17x + 4\sin 2x - \frac{15}{2}\cos 2x + c$ **ii** $\pi(\frac{17}{4}\pi + \frac{23}{2})$
   c i See CD.   **ii** 1.98^c (2 d.p.), 2.36^c (2 d.p.)
17 See CD.

SKILLS CHECK 5A (page 42)

1 **a** 2 **b** $\log_{10} 2$ **c** 16
2 **a i** $\dfrac{dx}{dt} = -15e^{-5t}$ **ii** $\dfrac{dx}{dt} = -5x$
  b $-15e^{-20}$ **c** 20
3 **a** 7389 **b** See graph on page 39.
  c i $14\,778$/h
      ii Draw the tangent at $t = 1$. It is the gradient of this tangent.
4 **a** 128 (3 s.f.)   **b** 5.16 per hour (3 s.f.)
5 **a** £1568.31   **b** 8 years
6 **a** 50 **b** 0.0023 (2 s.f.)
  c 0.065 grams per year (2 s.f.)

Exam practice 5 (page 43)

1 **a** £100   **b** £121.55   **c** 8.31 (3 s.f.)
2 **a** Substitute $t = 7$ **c** 2017
3 **a i** 50 **ii** 100 **b** 2.8 (2 s.f.)
4 **a i** 20 **ii** 90 **b** 22 months
5 **a i** 84 **ii** 21 **c** 13.7 (3 s.f.)
6 **a i** 50 cm
    ii 3 years 2 months
    iii 7.16 cm/year
  b As $x \to \infty$, $y \to 130$. A teenager's height is likely to be greater than 130 cm.
7 **a** See CD.   **b** 4.54 (3 s.f.)

SKILLS CHECK 6A (page 49)

1 **a** $y = Ae^{\sin x}$ **b** $y = (x^2 + c)^2$ **c** $y = Ax$
2 $t = \ln(x^2 - 3)$
3 **a** $\ln|y + 1| = \frac{1}{2}x^2 + c$ **b** $y = 3e^{\frac{1}{2}x^2} - 1$
4 **a** $-\dfrac{1}{y^2} = \frac{1}{3}x^3 + c$
5 $x^2 + y^2 - y - \frac{15}{4} = 0$, centre $(0, \frac{1}{2})$, radius 2
6 **a** $x = 4t^{\frac{1}{4}} - 3$ **b** $x = 4e^{\frac{1}{4}(t-1)} - 3$
7 $\tan\theta = \frac{1}{2}x^2 + c$
8 **a** $\frac{1}{2}x\sin 2x + \frac{1}{4}\cos 2x + c$
  b i $\frac{1}{2}(1 + \cos 2x)$ **ii** $\frac{1}{2}x + \frac{1}{4}\sin 2x$
  c $\frac{1}{2}y + \frac{1}{4}\sin 2y = \frac{1}{2}x\sin 2x + \frac{1}{4}\cos 2x - \dfrac{\pi}{8}$
9 **a** $\frac{1}{15}(9 + x^3)^5 + c$ **b** $y = \ln(9 + x^3)^5$
10 **a, b, c** See CD.
   d $56.7\,°$C (1 d.p.)

SKILLS CHECK 6B (page 54)

1 **a** $\dfrac{6x + 5}{2y + 6}$ **b** $\dfrac{2 - 2xy}{3y^2 + x^2}$ **c** $\dfrac{6xy - y\ln y}{3y^3 + x}$
2 $(3, 3), (3, 5)$
3 **a** $\dfrac{x - 2}{3 - 3y}$ **b** $y + 2x = 9$
4 **a** $-2, \frac{1}{2}$ **b** $90°$
5 **a** $\dfrac{1}{t}$ **b** $\frac{1}{2}\operatorname{cosec} t$ **c** $\dfrac{\ln 4t + 1}{2t - 4}$
6 **a** $\dfrac{2}{t}$ **b** $2y + x = 18$ **c** $p = 18, q = 9$
8 $a = 4$
9 $(1, 1)$

SKILLS CHECK 6C (page 57)

1 **a** $A = 2, B = -1$ **b** $\frac{2}{3}\ln|3x+1| - \frac{1}{2}\ln|2x-5| + c$
2 **a** $\frac{1}{x+2} - \frac{4}{2x-1} + \frac{2}{x+3}$ **b** $\ln\frac{81}{2}$
3 **a** $A = 1, B = -\frac{3}{2}, C = \frac{3}{2}$ **b** $x - \frac{3}{2}\ln|x+3| + \frac{3}{2}\ln|x-3| + c$
4 **a** $\frac{1}{x} - \frac{1}{x-1} + \frac{1}{(x-1)^2}$ **b** $\ln\frac{3}{4} + \frac{1}{2}$
5 **a** $A = 1, B = 4, C = 1$ **b** 15.3 (3 s.f.)

Exam practice 6 (page 58)

1 **a** $y = \sqrt[3]{3x + c}$ **b** $y = \sqrt[3]{3x - 4}$
2 $y = -\ln(1 + e - e^x)$
3 **a** $xe^x - e^x + c$ **b** $\ln y = xe^x - e^x + 1$
4 **a** Substitute $T = 7$ **b** 1.23 p.m.
5 **b** $\frac{1}{6}$
6 **b** 1.8 (2 s.f.)
7 **b** **i** $P = 1000e^{kt}, k = \frac{1}{30}\ln 2$ **ii** 22
8 130 cm
   b **i** $\frac{1}{2}r^2 = kt + c$ **ii** 162.8 cm (1 d.p.) **iii** See CD.
9 **a** **i** $\frac{1}{6}$ **b** $y^2 = \ln(x+2) + 1 - \ln 3$
10 **a** $\pm\frac{5\sqrt{5}}{3}$ **b** ± 1.5 (2 s.f.)
11 **a** $(2, -1), (2, -3)$ **b** $\frac{4}{9}, -\frac{4}{9}$
12 **b** $y - 1 = -\frac{4}{3}(x - 1)$
13 **a** $-\frac{1}{t}$ **b** $y = 3x + 30$
14 **b** $y = -\frac{1}{3}x + 1.414$
15 **a** **i** $\frac{1}{t}$ **ii** 2 **b** **i** $x = \frac{y^2}{12}$ **ii** $\frac{dx}{dy} = \frac{y}{6}$
16 **a** $\frac{2}{x+4} + \frac{4}{7-2x}$ **b** $4\ln(\frac{7}{2})$
17 **a** $\frac{-3}{x+4} + \frac{4}{2x+1}$ **b** $4\ln 3 - 3\ln 2$
18 **a** **ii** $\frac{2}{x-4} - \frac{2}{x+4}$ **b** $3 + 2\ln 3$
19 **a** $\frac{1}{2x} + \frac{2}{x-1} - \frac{1}{(x-1)^2}$ **b** $\frac{1}{2}\ln|x| + 2\ln|x-1| + \frac{1}{x-1} + c$
20 **c** **i** $y - 3 = -\frac{3}{5}(x - 5)$
    d **i** $x + y = 8t$
       $x - y = \frac{2}{t}$
    ii $x^2 - y^2 = 16$
    e $7.5 - 8\ln 2$

SKILLS CHECK 7A (page 68)

1 **a** Neither **b** Equal **c** Parallel
2 $\begin{bmatrix} 4 \\ 5 \\ -6 \end{bmatrix}$
3 **a** 13 **b** $\sqrt{14}$ **c** 5
   d $2\sqrt{5}$ **e** $\sqrt{30}$ **f** $\sqrt{6}$
4 **a** $\begin{bmatrix} 7 \\ -5 \\ 9 \end{bmatrix}$ **b** $\begin{bmatrix} 3 \\ 5 \\ -3 \end{bmatrix}$ **c** $\begin{bmatrix} 4 \\ 1 \\ -6 \end{bmatrix}$
   d $\sqrt{17}$ **e** $\sqrt{61}$ **f** $\sqrt{34}$
5 **a** $\begin{bmatrix} -7 \\ -8 \\ -7 \end{bmatrix}$ **b** $\begin{bmatrix} 6 \\ 5 \\ 3 \end{bmatrix}$ **c** $\begin{bmatrix} 7 \\ 2 \\ -2 \end{bmatrix}$
   d $\sqrt{38}$ **e** $3\sqrt{17}$ **f** $3\sqrt{11}$

6 $\begin{bmatrix} \frac{\sqrt{3}}{15} \\ \frac{\sqrt{3}}{3} \\ \frac{-7\sqrt{3}}{15} \end{bmatrix}$

7 **a** **i** $\begin{bmatrix} 2 \\ -3 \end{bmatrix}$ **ii** $\begin{bmatrix} 3 \\ 2 \end{bmatrix}$ **iii** $\begin{bmatrix} 1 \\ 5 \end{bmatrix}$
   b **i** $\sqrt{13}$ **ii** $\sqrt{13}$ **iii** $\sqrt{26}$
8 $-1, 3$

SKILLS CHECK 7B (page 70)

1 **a** $\begin{bmatrix} -4 \\ -5 \\ 2 \end{bmatrix}$ **b** $\begin{bmatrix} 2 \\ -4 \\ 6 \end{bmatrix}$ **c** $\begin{bmatrix} -3 \\ 5 \\ -2 \end{bmatrix}$
2 **a** **i** $2\mathbf{b} - \mathbf{a}$ **ii** $-3\mathbf{a} + (k-3)\mathbf{b}$
   b 9 **c** $1 : 3$
3 **a** 7.3 **b** 9.2 **c** 17.1
4 **a** $|\overrightarrow{AB}| = |\overrightarrow{BC}| = \sqrt{18}$ **b** $\begin{bmatrix} 7 \\ -1 \\ 3 \end{bmatrix}$
5 $-1, 3$

SKILLS CHECK 7C (page 74)

Note that equivalent forms of the vector equation of a line would be acceptable.

1 **a** $\mathbf{r} = \begin{bmatrix} 2 \\ 0 \\ -1 \end{bmatrix} + \lambda \begin{bmatrix} 7 \\ -2 \\ 6 \end{bmatrix}$ **b** $\mathbf{r} = \begin{bmatrix} 4 \\ -1 \\ 3 \end{bmatrix} + \lambda \begin{bmatrix} 0 \\ 6 \\ -1 \end{bmatrix}$
   c $\mathbf{r} = \begin{bmatrix} -1 \\ 2 \\ 1 \end{bmatrix} + \lambda \begin{bmatrix} 1 \\ 1 \\ 0 \end{bmatrix}$
2 **a** $\mathbf{r} = \begin{bmatrix} 2 \\ -1 \\ 5 \end{bmatrix} + \lambda \begin{bmatrix} -5 \\ 1 \\ -4 \end{bmatrix}$ **b** $\mathbf{r} = \begin{bmatrix} 0 \\ 2 \\ 1 \end{bmatrix} + \lambda \begin{bmatrix} 3 \\ 1 \\ -2 \end{bmatrix}$
   c $\mathbf{r} = \begin{bmatrix} 1 \\ 4 \\ -2 \end{bmatrix} + \lambda \begin{bmatrix} -4 \\ -3 \\ 6 \end{bmatrix}$
3 $t = -1$
4 $a = 4, b = -2$
6 **a** $(-1, 4, -7)$ **b** $(2, -3, 7)$
7 **b** $5\sqrt{14}$
8 **a** $\mathbf{r} = \begin{bmatrix} 2 \\ 1 \\ -2 \end{bmatrix} + \lambda \begin{bmatrix} -5 \\ 3 \\ 3 \end{bmatrix}$ **b** $p = 22, q = -14$

SKILLS CHECK 7D (page 79)

1 **a** $106.9°$ (1 d.p.) **b** $90°$ **c** $75.6°$ (1 d.p.)
2 $\mathbf{a}.\mathbf{b} = 12 - 7 - 5 = 0$
3 9.21
4 **a** $40.4°$ (1 d.p.) **b** $17.3°$ (1 d.p.)
5 $\sqrt{\frac{13}{3}}$
6 $\angle BAC = 81.0°, \angle ACB = 37.6°, \angle ABC = 61.5°$
7 **b** $13\sqrt{3}$
8 **b** $(1, 5, 2)$
9 $1, 1\frac{1}{2}$
10 $(\frac{11}{9}, -\frac{26}{9}, -\frac{2}{9})$
11 **a** $A(5, 2, 9), AP = \sqrt{62}$ **b** **i** $-\frac{1}{5}$ **ii** 3.41 (3 s.f.)
12 **c** $(3, 1, 3)$ **d** $\sqrt{3762}$

Exam practice 7 (page 80)

Equivalent forms of the vector equation of a line are acceptable.

1. **a** 2   **b** $\begin{bmatrix} 4 \\ 7 \\ -5 \end{bmatrix}$

2. **b** 77° (nearest °)

3. **a** $\mathbf{r} = \begin{bmatrix} 3 \\ -1 \\ 2 \end{bmatrix} + \lambda \begin{bmatrix} -1 \\ 1 \\ 0 \end{bmatrix}$
   b (1, 1, 2)   **d** (5, −3, 2)

4. **a** (2, 2, −2)   **b** 71° (nearest degree)

5. **a** $a = 1, b = 11$ **b** (7, 2, 4)   **c** 54.2°

6. **b i** $t = -1$

7. **b** 71.6°

8. **a**

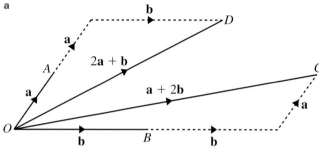

   b $\overrightarrow{AB} = \mathbf{b} - \mathbf{a}$, $\overrightarrow{BC} = \mathbf{b} + \mathbf{a}$, $\overrightarrow{DC} = \mathbf{b} - \mathbf{a}$, $\overrightarrow{AD} = \mathbf{b} + \mathbf{a}$

9. **b** 21.1° (3 s.f.)

10. **a** $\mathbf{r} = \begin{bmatrix} 2 \\ 1 \\ -3 \end{bmatrix} + \lambda \begin{bmatrix} 1 \\ -1 \\ 1 \end{bmatrix}$
    c 84° (nearest °)

11. **a** $\mathbf{r} = \begin{bmatrix} 0 \\ 4 \\ -2 \end{bmatrix} + \lambda \begin{bmatrix} 1 \\ -2 \\ 3 \end{bmatrix}$ **b** (1, 2, 1)   **c** $\sqrt{21}$

Practice exam paper (page 84)

1. **a** $\sqrt{104} \cos(x + 11.3°)$ **b** 49.3°, 288.0°
   c $-\sqrt{104}$

2. **a** $\dfrac{dy}{dx} = -\dfrac{2}{t^2}$ **b** e.g. $y - 2 = 8(x - 11)$
   c $(x - 3)(y - 1) = 8$

3. **a** $2\frac{2}{9}$ **b** 0   **c** $\dfrac{1}{2x - 1}$

4. **a** $\dfrac{3}{x + 2} + \dfrac{4}{2x - 3}$ **b** $3\ln|x + 2| + 2\ln|2x - 3| + c$

5. **a** $1 + \frac{1}{4}x - \frac{3}{32}x^2 + \ldots$

6. **a** $\cos^2 x - \sin^2 x$ **c** $\dfrac{\pi}{8}$

7. **a** $100 \ln \frac{4}{3}$ **b** $y^2 = \dfrac{1}{x^2 - 4x + 5}$

8. **a** $\begin{bmatrix} 2 \\ 1 \\ 3 \end{bmatrix}$ **c ii** $3\sqrt{2}$

SINGLE USER LICENCE AGREEMENT FOR CORE 4 FOR AQA CD-ROM
IMPORTANT: READ CAREFULLY

WARNING: BY OPENING THE PACKAGE YOU AGREE TO BE BOUND BY THE TERMS OF THE LICENCE AGREEMENT BELOW.

This is a legally binding agreement between You (the user or purchaser) and Pearson Education Limited. By retaining this licence, any software media or accompanying written materials or carrying out any of the permitted activities You agree to be bound by the terms of the licence agreement below.

If You do not agree to these terms then promptly return the entire publication (this licence and all software, written materials, packaging and any other components received with it) with Your sales receipt to Your supplier for a full refund.

YOU ARE PERMITTED TO:

- Use (load into temporary memory or permanent storage) a single copy of the software on only one computer at a time. If this computer is linked to a network then the software may only be used in a manner such that it is not accessible to other machines on the network.
- Transfer the software from one computer to another provided that you only use it on one computer at a time.
- Print a single copy of any PDF file from the CD-ROM for the sole use of the user.

YOU MAY NOT:

- Rent or lease the software or any part of the publication.
- Copy any part of the documentation, except where specifically indicated otherwise.
- Make copies of the software, other than for backup purposes.
- Reverse engineer, decompile or disassemble the software.
- Use the software on more than one computer at a time.
- Install the software on any networked computer in a way that could allow access to it from more than one machine on the network.
- Use the software in any way not specified above without the prior written consent of Pearson Education Limited.
- Print off multiple copies of any PDF file.

ONE COPY ONLY

This licence is for a single user copy of the software

PEARSON EDUCATION LIMITED RESERVES THE RIGHT TO TERMINATE THIS LICENCE BY WRITTEN NOTICE AND TO TAKE ACTION TO RECOVER ANY DAMAGES SUFFERED BY PEARSON EDUCATION LIMITED IF YOU BREACH ANY PROVISION OF THIS AGREEMENT.

Pearson Education Limited and/or its licensors own the software.
You only own the disk on which the software is supplied.

Pearson Education Limited warrants that the diskette or CD-ROM on which the software is supplied is free from defects in materials and workmanship under normal use for ninety (90) days from the date You receive it. This warranty is limited to You and is not transferable. Pearson Education Limited does not warrant that the functions of the software meet Your requirements or that the media is compatible with any computer system on which it is used or that the operation of the software will be unlimited or error free.

You assume responsibility for selecting the software to achieve Your intended results and for the installation of, the use of and the results obtained from the software. The entire liability of Pearson Education Limited and its suppliers and your only remedy shall be replacement free of charge of the components that do not meet this warranty.

This limited warranty is void if any damage has resulted from accident, abuse, misapplication, service or modification by someone other than Pearson Education Limited. In no event shall Pearson Education Limited or its suppliers be liable for any damages whatsoever arising out of installation of the software, even if advised of the possibility of such damages. Pearson Education Limited will not be liable for any loss or damage of any nature suffered by any party as a result of reliance upon or reproduction of or any errors in the content of the publication.

Pearson Education Limited does not limit its liability for death or personal injury caused by its negligence.

This licence agreement shall be governed by and interpreted and construed in accordance with English law.